Vineyard Soil

-Selected Articles-

by

Various Authors

Copyright © 2013 Read Books Ltd.
This book is copyright and may not be
reproduced or copied in any way without
the express permission of the publisher in writing

British Library Cataloguing-in-Publication Data
A catalogue record for this book is available from the
British Library

Winemaking

The science of wine and winemaking is known as 'oenology', and winemaking, or 'vinification', is the production of wine, starting with selection of the grapes or other produce and ending with bottling the finished product. Although most wine is made from grapes, it may also be made from other fruits, vegetables or plants. Mead, for example, is a wine that is made with honey being the primary ingredient after water and sometimes grain mash, flavoured with spices, fruit or hops dependent on local traditions. Potato wine, rice wine and rhubarb wines are also popular varieties. However, grapes are by far the most common ingredient.

First cultivated in the Near East, the grapevine and the alcoholic beverage produced from fermenting its juice were important to Mesopotamia, Israel, and Egypt and essential aspects of Phoenician, Greek, and Roman civilization. Many of the major wine-producing regions of Western Europe and the Mediterranean were first established during antiquity as great plantations, and it was the Romans who really refined the winemaking process.

Today, wine usually goes through a double process of fermentation. After the grapes are harvested, they are prepared for primary fermentation in a winery, and it is at this stage that red wine making diverges from white wine making. Red wine is made from the must (pulp) of red or black grapes and fermentation occurs together with the grape skins, which give the wine its colour. White wine is made by fermenting juice which is made by pressing crushed grapes to extract a juice; the skins are removed and play no further role. Occasionally white wine is made from red grapes; this is done by extracting their juice with minimal contact with the grapes' skins. Rosé wines are either made from red grapes where the juice is allowed to stay in contact with the dark skins long enough to pick up a pinkish colour (blanc de noir) or by blending red wine and white wine.

In order to embark on the primary fermentation process, yeast may be added to the must for red wine or may occur naturally as ambient yeast on the grapes or in the air. During this fermentation, which often takes between one and two weeks, the yeast converts most of the sugars in the grape juice into ethanol (alcohol) and carbon dioxide. The next process in the making of red wine is secondary fermentation. This is a bacterial fermentation which converts malic acid to lactic acid, thereby decreasing the acid in the wine and softening the taste. Red (and sometimes White) wine is sometimes transferred to oak barrels to mature for a period of weeks or months; a practice which imparts oak aromas to the wine. The end product has been both revered as a highly desirous and delicious status symbol, as well as a mass-produced, cheap form of alcohol.

Interestingly, the altered consciousness produced by wine has been considered *religious* since its origin. The Greeks worshipped Dionysus, the god of winemaking (as well as ritual madness and ecstasy!) and the Romans carried on his cult under the name of Bacchus. Consumption of ritual wine has been a part of Jewish practice since Biblical times and, as part of the Eucharist commemorating Jesus' Last Supper, became even more essential to the Christian Church.

Its importance in the current day, for imbibing, cooking, social and religious purposes, continues. Winemaking itself, especially that on a smaller scale is also experiencing a renaissance, with farmers and individuals alike re-discovering its joy.

Contents

Classification of Soils ....*page 1*
William Hardman

Soil, Situation and Aspect ....*page 9*
John Phin

Preparation of the Soil ....*page 33*
John Phin

Soil and Cultivation ....*page 63*
William C. McCollom

Location and Soil, Preparation of the Ground
    And How to Cultivate the Soil ....*page 71*
Charles Reemelin

The Soil and its Preparation ....*page 89*
Peter B. Mead

Soil and Situation ....*page 101*
William Chamberlain Strong

Soil and Situation ....*page 123*
Andrew S. Fuller

Classification of Soils

by

William Hardman

Classification of Soils.

It will be well at this stage to describe the mode of testing soils. Spread a weighed quantity of the soil in a thin layer upon writing paper, and dry it for an hour or two in an oven, or upon a hot plate, the heat of which is not sufficient to discolour the paper; the loss of weight gives the water it contained. While this soil is drying, a second weighed portion may be boiled or otherwise thoroughly incorporated with water, and the whole then poured into a vessel, in which the heavy sandy parts are allowed to subside until the fine clay is beginning to settle also; this point must be carefully watched, the liquid then poured off, the sand collected, dried as before upon the paper, and again weighed. This weight is the quantity of sand in the known weight of moist soil, which, by the previous experiment, has been found to contain a certain quantity of water.

Thus, suppose two portions, each 200 grains, are weighed, and the one in the oven loses 50 of water, and the other leaves 60 grains of sand, then the 200 grains of moist are equal to 150 grains of dry soil, and contain 60 of sand or 40 in 100 (40 per cent.); it would therefore be properly called a clay loam.

Marly soils are those in which the proportion of lime is more than 5, but does not exceed 20 per cent. of the whole weight of the dry soil. The marl is a sandy, loamy, or clay marl, according as the proportion of clay it contains would place it under the one or the other denomination, supposing it to be entirely free from lime, or not to contain more than 5 per cent.

Calcareous soils, in which the lime, exceeding 20 per cent., becomes the distinguishing constituent; these are also called calcareous clays, loams, or sands, according to the proportion of pure clay which is present in them. The determination of the lime when it exceeds 5 per cent. is attended with no difficulty. Thus, to 100 grains of the dry soil, diffused through half a pint of cold water, add half a wine-glassful of muriatic acid (the spirit of salt of the shops), stir it occasionally during the day, and let it stand over night to settle. Pour off the clear liquor in the morning, and fill up the vessel with water to wash away the excess of acid; when the water is again clear, pour it off, dry the soil, and weigh it; the loss will amount generally to one per cent. more than the quantity of lime present; the result will be sufficiently near, however, for the purposes of classification. If the loss exceeds 5 grains from 100 of the dry soil, it may be classed among the marl; if more than 20 grains, among the calcareous soils.

The several steps, therefore, which are necessary in examining a soil, with a view of so far determining its composition as to be able precisely to name and classify it, will be best taken in the following order:—Weigh 100 grains of the soil, spread them in a thin layer upon white paper, and place them for some hours in an oven or other hot place, the heat of which may be raised till it begins slightly to discolour the paper. The loss is water. Secondly. Let it now (after drying and weighing) be again placed over the fire, heat it to dull redness, or over a lamp, till the combustible matter is burned away; the second loss is organic, chiefly vegetable matter with a little water which still remained in the soil after drying. Thirdly. After being thus burned, let it be put into half a pint of water, with half a wine-glassful of spirit of salt, and frequently stirred when minute bubbles of air cease to rise from the soil. On settling, this process may be considered as at an end. The loss by this treatment will be a little more than the true per-centage of lime, and it will generally be nearer the truth if that portion of the soil be employed which has been previously heated to redness. Fourthly. A fresh portion of the soil, perhaps 200 grains, in its moist state, may now be taken and washed, to determine the quantity of silicious sand it contains. If the residual sand be supposed to contain calcareous matter, its amount may readily be determined by treating the dried sand with diluted muriatic acid in the same way as when determining the whole amount of lime (3°) contained in the unwashed soil.

The weighings for the purposes here described may be made in a small balance, with grain weights, sold by a druggist for 5s. or 6s., and the vegetable matter may be burned away on a slip of sheet iron, or in an untinned iron table-spoon, over a bright cinder or charcoal fire, care being taken that no scale of oxide which may be formed in the iron be allowed to mix with the soil when cold, and thus increase its weight. Those who are inclined to perform the latter operation more neatly may obtain for about 6s. each, from the dealers in chemical apparatus, thin, light, platinum capsules, from 1 to 1½ inches in diameter, capable of holding 50 or 100 grains of soil, and for a few shillings more a spirit-lamp, over which the vegetable matter of the soil may be burned away. With care, one of these capsules will serve a lifetime.

Example—Selected a portion of soil.

1. After being dried in the air, and by keeping some time in paper, it was dried for some hours at a temperature sufficient to give the white paper below it a scarcely perceptible tinge; by this process 104½ grains lost 4 grains.

2. When thus dried it was heated to dull redness, it first blackened, but gradually assumed a pale brick colour, the change of course beginning at the edges; the loss by this process was 4½ grains.

3. After this heating it was put into half a pint of pure rain water, with half a wine-glassful of spirit of salt; after some hours, when the action had ceased, the soil was washed and dried again at a dull red heat; the loss amounted to three grains.

4. Washed with water by decantation 100 grains of the soil, left 70 grains of very fine sand, or 104½ would have left 73 grains.

The soil therefore contained:—

	Grains.	
Water	4	3.9 per cent.
Organic matter less than	4⅓	4.1
Carbonate of Lime	8	3.0
Clay	20¼	19.0
Sand, very fine	73	70.0
	104½	100.0

The soil, therefore, containing 70 per cent. of sand, separable by decantation, is properly a sandy loam.

A loamy soil deposits 40 to 70 per cent. of sand by mechanical washing.

A sandy loam leaves from 70 to 90 per cent. of sand.

A sandy soil contains no more than 10 per cent. of pure c'ay.

Clay soils are considered best for wheat, loamy for barley; sandy loams, for oats or barley; such as are more sandy still, for oats and rye; or those almost pure sand, for rye alone.

Soil, Situation and Aspect

by

John Phin

SOIL, SITUATION AND ASPECT.

Soil.—The vine will grow in almost any situation, and reach a large size and exhibit luxurious vegetation under conditions apparently the most unfavorable; but if healthy vines and fine fruit be desired, it is necessary to choose a soil where the roots can ramble freely, find plenty of nutriment and be safe from stagnant water and its accompanying cold, sour subsoil. One of the largest vines in the country grows in a swamp in New Jersey, and a vine has been known to grow vigorously from a cleft in an old wall twenty feet from the ground. But these are by no means examples to be imitated in practice where we have the power of selecting the site of our garden or vineyard, though they afford encouragement to the amateur who is compelled to make use of an inferior location.

The opinion of good grape culturists is that any soil which will grow *good* Indian corn is suitable for grapes. Others describe a soil adapted to the culture of the vine as one which will grow good winter

SOIL, SITUATION AND ASPECT.

wheat without the plants being thrown out of the ground in winter.

Downing recommends a "strong loamy or gravelly soil—limestone soils being usually the best." And in another place he gives it as his opinion that "all that can be said of a soil for grape culture is that it be light, rich and dry." G. W. Johnson thinks a light, sandy loam the best. And Buchannan, who may be safely taken as the representative of the Cincinnati vine growers, recommends a dry, calcareous loam with a porous subsoil. At the recent meeting of the Fruit Growers' Society of western New York, Dr. Farley stated that his best grapes had been raised on a clay soil, and that in this matter his opinion in regard to the soil best adapted to the culture of grapes had undergone some change.

It will thus be perceived that the opinions of our best horticulturists vary a little, but we believe that this variation is mere adaptation to the different modes of growth and training adopted by the various cultivators. The purpose for which the grapes are raised —that is whether for wine or for the table—ought also to have a material influence in directing our choice of a soil.

When the object is to manufacture wine, the vines require to be kept within moderate bounds; all rankness of vegetation must be carefully avoided, and consequently the soil must be light, rich, porous and dry, and if calcareous so much the better.

On the other hand, where high saccharine qualities are not so much desired as abundance of grapes of agreeable flavor, the vines will succeed better and produce more certain crops if allowed a greater extent of growth, and in this case they will bear a heavier and richer soil—in some cases (as in growing Isabella and Diana grapes for the table) even preferring a clay soil well drained and cultivated and highly manured.

That this view is correct may be easily proved by referring to well-known examples both in Europe and in this country. Thus in the Arriege in France a rich wine like Tokay, is obtained from mountain sides covered with large stones as if the cultivators had left all to nature. In Italy and Sicily the best wines are grown amongst the rubbish of volcanoes. "Good rich soils," says Redding, "never produce even tolerable wines."

On the other hand the rich Chasselas de Fontainebleau table grapes are produced by vines planted in cold and heavy soil, well manured. And he who desires to find *rich soil* should examine the vine borders of the English hot-house grape-growers. Allen, one of our most successful grape-growers recommends a border of the richest kind. So does Chorlton, and such we believe to be the practice of all our successful cultivators of the grape under glass. The celebrated vine at Hampton Court revels in the luxury of an old sewer, and instances have come under our own observation where the proximity of a vine to a cesspool caused the production of large quantities of most excellent grapes. In France, the application of night-soil and sewerage to the vineyards has in all cases injured the quality of the wine. That such would have been the case, however, if the French vignerons had acted upon correct principles in the application of these powerful stimulants, we are scarcely prepared to believe. And we have no doubt but that by judicious management and a careful observance of the laws of nature one of the greatest achievements in vine culture may yet be effected, viz., the union of vigorous vegetation and stimulating manures with the production of good wine. But so far as present experience extends the soil for a vineyard must be light and not *too highly manured*—and in all cases whether the object of culture be wine or table grapes the subsoil must be warm and loose. Cold borders are very prejudicial to the roots of the vine, and are supposed to be an efficient cause of the *shanking* of the grapes. It would appear from an inspection of the subjoined tables that this desired warmth might be secured to the surface soil at least by plentiful addition of lime and any black mold or charcoal.

Maximum Temperatures of the various Earths Exposed to the Sun. By Schubler.

KINDS OF EARTH.	Maximum Temperature of the superior layer, the mean temperature of the ambient air being 77 degrees F.	
	Moist Earth.	Dry Earth.
	Degrees.	Degrees.
Silicious sand, yellowish grey, ...	99.05	112.55
Calcareous sand, whitish grey,....	99.10	112.10
Argillaceous earth, yellowish grey,	99.28	112.32
Calcareous earth, white, ...	96.13	109.40
Mold, blackish grey,............	103.55	117.27
Garden earth, blackish grey,.....	99.50	113.45

Table of Retention of Heat. By Becquerel.

KIND OF EARTH.	Capacity for heat, that of Calcareous sand being 100.	Time required by 18 feet cube of earth to cool from 144.5 to 70.2, the temperature of the surrounding air being 61°.2.
		Hours.
Calcareous sand, ...	100	3.30
Silicious sand,........	95.6	3.27
Argillaceous earth,....	68.4	2.24
Calcareous earth,.....	61.8	2.10
Mold, ...	49	1.43

From these tables it will be seen that black mold receives or absorbs heat most rapidly, but parts with it in the shortest space of time also, and that for receiving and retaining heat, dark colored, calcareous earth is by far the most efficient. Good silicious sand comes next in order, and hence we conceive that a soil composed chiefly of calcareous and silicious sand, with a sufficient amount of charcoal or mold to give it a dark color, would prove one of the best for grapes.

Such are the general points deserving of consideration. Those desirous of studying more minutely the influence of the chemical constitution of the soil upon vines growing therein will find an interesting and valuable résumé of the subject in M. Ladrey's "Chimie appliqué à la Viticulture," whose general remarks on this point are so much in unison with our own experience and observation that we are tempted to translate them.

"If now we examine the series of different soils devoted to the culture of the vine in France and in other countries, we shall find this plant cultivated in soils the most diverse, not only as regards their natures (nature evidently alluding to physical constitution—*Trans.*)—but also their chemical composition. All soils appear suited to the culture of the vine, and there are none, unless those absolutely barren, in which this plant may not grow and develop itself. Thus the vine requires but little fertility in the soil, it covers a great space of land which would be unsuited to any other culture, and in order to give an idea of this, we may cite the ancient regulations of Provence which prohibited the planting of the vine until inquiry had been made as to the sterility of the soil, and the permission of the intendant of the province had been obtained.

But if the vine can grow in all soils it behaves very differently in each of them. In strong, argillaceous, rich soils, it will acquire a great vigor of vegetation, the wood is largely developed, the product is abundant; on the contrary, in soils poor, light and dry, the vine is less robust, more delicate; it requires a culture well contrived as to even the most minute details, and the product is much less in quantity.

"In general, if in any locality the vegetation of the vine be more rich as the soil is more fertile, we observe by the side of this result that the nature and quality of the product is consequently in an inverse ratio. In heavy land the vine is well developed and furnishes abundant return; in a light soil it gives less and the product is of higher quality."

SITUATION.—THE situation of a vineyard should be elevated, but not too high, otherwise the vines will not only be exposed to high winds and their concomitant evils, but will also be subjected to a lower temperature. On this latter point, but little is known—at

SOIL, SITUATION AND ASPECT.

least not enough to enable us in all cases to reconcile the anomalies which occur. Enough is known, however, to cause us to avoid the tops of hills and the bottoms of valleys, and it may be worth our while to consider a few of the principles which regulate temperature in these situations. During the night, the cold air, being heavy, settles down into the valleys and hollows, thus producing in such locations a temperature several degrees lower than is found on the sides of the adjacent hills. And no influence is then at work to disturb this state of things, for the earth itself is becoming rapidly cooled by radiation; and if a small quantity of the air should become warmed by contact with it, it immediately ascends, and cool air takes its place.

At daybreak, however, an agency is introduced which reverses this condition of things. Then the dense air in the valleys concentrates and absorbs the heat of the sun's rays and increases their effect upon the soil, which in turn imparts heat to the stratum of air lying next it. This lower stratum of air being warmed and consequently rendered much lighter than the colder portion above it, it ascends, but as it rises it also expands still more, which in some measure compensates for the heat which it received from the earth. The same process keeps going on until night comes, when the lower stratum of air being no longer warmed it no longer ascends, and the colder and heavier air again accumulates in the valleys. Thus it will be seen, that during the night the air in the valleys is colder than that in other places, while the reverse is the case during the day. The stillness of the air in valleys and sheltered situations also contributes to this result in a remarkable degree.

Now it is obvious, that if for any fruit tree, the air in the valleys should be sufficiently cold to kill the buds, no orchard could succeed. And if, on the other hand, sufficient light and heat to ripen the fuit could not be found on the hill-tops, such situations also would be unavailable.

Nor is the mere existence of such extremes of temperature the worst evil. The destructive influence of a hot sun upon frozen vegetation is well known, and in low valleys, the circumstances are such as to give the greatest effect to this adverse influence. For not only are the plants chilled by the extra cold night-air, they are also completely protected from the rays of the sun, until it has attained a greater power than it usually exerts at its first appearance upon plants in more exposed situations. And then, owing to the dense atmosphere through which they pass, the rays strike suddenly with concentrated energy so as to thaw the buds with a rapidity completely destructive to their vitality. In such situations also, the soil is usually very deep and rich, producing a vigorous though succulent growth which is unable to withstand the influences above detailed. All experience bears out the practical value of these principles. Thus, in Italy, where the country is undulating and very much broken, all good wines are grown on the hill-sides. Hence Virgil tells us

. . . . . "denique apertos
Bacchus amat colles,"*

and modern experience bears out the ancient saw, though it does not follow, however, that plains will not produce good wine-making grapes, provided they be of sufficient extent to obviate the evils just described. The fine wines of the Gironde in France, and Châtaux Margaux, Lafitte and Latour, are grown on the plains.

ASPECT—EXPOSURE.—The aspect which is best adapted to the growth of grapes will, of course, depend upon influences, some of which at least, are liable to vary, as the keenest and most destructive winds may come from different quarters in different places—a very slight geographical change sometimes making

* The force of this saying is lost by adopting Mr. Redding's translation "Bacchus loves the hills." Davidson gives the whole, " Bacchus loves the open hills"—which is better. But the true meaning " Bacchus loves the open little hills " coincides perfectly with experience and with the principles above set forth.

an important difference in this respect, owing to peculiar topographical features. Thus a range of hills or a belt of woods, may so deflect the prevailing winds, as to completely change the condition of two localities situated within even a very short distance of each other.

In general, it will be found necessary to secure protection on the west, north and northeast. This may be afforded either by natural local features, as by a range of hills, or it may be derived from artificial sources, as woods or fences. No defence is better than a good belt of Norway spruce, and if they form a crescent in which the vineyard is embowered, but little danger need be apprehended from violent winds. Even high fences, which may be single, double or triple, afford ample protection in ordinary cases, and as trees, even of the fastest growing kind, take a considerable time before they give sufficient protection, many will prefer the fence. We are therefore tempted to extract from the "Horticulturist" for August, 1847, Downing's description of the method by which Frederic Tudor, Esq., has converted the naked promontory of Nahant into a luxuriant garden.

"To appreciate the difficulties with which this gentleman had to contend, or as we might more properly say, which stimulated all his efforts, we must recall to mind that, frequently, in high winds,

SOIL, SITUATION AND ASPECT.

the salt spray drives over the whole of Nahant; that until Mr. Tudor began his improvements, not even a bush grew naturally on the whole of its area; and that the east winds which blew from the Atlantic in the spring are sufficient to render all gardening possibilities in the usual way nearly as chimerical as cultivating the volcanoes of the moon. Mr. Tudor's residence there, now, is a curious and striking illustration of the triumph of art over nature.

"Of course, even the idea of a place worthy of the name of a garden in this bald, sea girt cape, was out of the question, unless some mode of overcoming the violence of the gales and the bad effect of the salt spray could be devised. The plan Mr. Tudor has adopted is, we believe, original with him, and is at once extremely simple and perfectly effective.

* * * * * * * *

"It consists merely of two, or at most three parallel rows of high open fences, made of rough slats or palings, nailed in the common vertical manner, about three inches wide, and a space of a couple of inches left between them. These paling fences are about 16 feet high, and usually form a double row (on the most exposed side, a triple row) round the whole garden. The distance between that on the outer boundary and the next interior one is about four feet. The garden is also intersected here and there by tall trellis fences of the same kind, all of which help to increase the shelter, while some of those in the interior serve as frames for training trees upon.

"The effect of this double or triple barrier of high paling is marvellous; although like a common paling, apparently open and permitting the wind a free passage, yet in practice it is found entirely to rob the gales of their violence and their saltness. To use Mr. Tudor's words, 'it completely sifts the air.' After great storms, when the outer barrier will be found covered with a coating of salt, the foliage in the garden is entirely uninjured. It acts, in short, like a rustic veil, that admits just so much of the air, and in such a manner as most to promote the growth of the trees, while it breaks and wards off all the deleterious influences of a genuine ocean breeze, so pernicious to tender leaves and shoots.'

* * * * * * * *

"It is worthy of record, among the results of Mr. Tudor's culture, that two years after the principal plantation of his fruit trees was made, he carried off the second prize for pears at the annual exhibition of the Massachusetts Horticultural Society, among dozens of zealous competitors, and with the fruit most carefully grown in that vicinity."

Of the necessity for shelter under circumstances far less desperate than those at Nahant, no good horticulturist has any doubt. Even the oak-tree has been proved by a well directed series of experiments, to be benefited by shelter in the comparatively mild climate of England. For the rationale of the evil effects of wind on plants in general, we must refer the reader to Lindley's "Theory and Practice of Horticulture." The following cases are detailed by Hoare :

"Many instances might be circumstantially detailed of the injurious effects of wind upon established vines during their summer's growth; two, however, of recent occurrence will perhaps suffice.

"On the eleventh of June, 1833, a strong wind sprang up early in the morning from the west, and increased in force till noon, when it blew quite a gale and continued to do so throughout the day. It slackened a little during the night, and gradually decreased in violence the next day, dying entirely away in the evening.

"The effects of this wind on a vine of the White Muscadine sort, trained on a wall having a western aspect, were carefully observed. It had on a full crop of fruit and a good supply of fine young bearing shoots, and was altogether in a most thriving condition. Such, however, were the injurious effects of the wind in dissipating all the accumulated secretions of the foliage, and then closing, almost hermetically, its pores, and thereby totally deranging the vital functions of the plant, that although in the height of the growing season, not the slightest appearance of renewed vegetation could be discerned in any part of its leaves, shoots or fruit, until the third day of July, or twenty-two days afterward. It never produced another inch of good bearing wood throughout the remainder of the season, but lingered in a very weak and sickly condition; and the fruit which had been previously estimated at ninety pounds' weight, did not exceed fifty-five pounds when gathered, and that of a very inferior description in point of flavor and size of berry. Its leaves, also, having been thus crippled, were shed prematurely a month before their natural time, and hence the deficiency in the flavor and size of the grapes.

"The other instance, which happened shortly afterward, is still more decisive. On the 30th of August following, about eight o'clock in the evening, a strong wind began to blow from the southwest, accompanied with heavy rain. At nine it blew violently, and continued to do so until noon the next day. It then slackened, and then veering to the northwest, died away some time during the following night.

"The full force of this wind fell on a remarkably fine black Hamburg vine, trained on a wall having a southwestern aspect, and its effects were therefore proportionately destructive. Many of the principal branches were torn so completely from their fastenings that their extremities swept the ground. The bunches of fruit were knocked about, and portions of them, as well as single berries, lay scattered on the ground in every direction. On the fruit, however, that survived the wreck, the effects of the wind were remarkable. It must be stated that the wall on which the vine is trained, is ten feet high, and is so situated that to the height of about three feet from the ground the wind had but little power over it, its force being broken by an outer wall standing at a little distance off in front of it. On the lower part of the wall so protected, the grapes not having been much injured, began to change their color and ripen about the twentieth of September, and on the twelfth of October every berry was perfectly matured, while all those that remained on the vine above three feet from the ground, were, on the first of November, as green and hard as on the thirtieth of August, when the high wind occurred. Shortly afterward these began to change their color, and ultimately ripened tolerably well by the first week in December. Thus, solely through the effects of a strong wind, there were to be seen at the same time, on the same branches of this vine, and within nine inches of each other, bunches of grapes, the lowermost of which were perfectly ripe, while the uppermost were quite green and hard, and not within seven weeks of reaching the same state of maturity.

"These facts, which might be multiplied indefinitely, sufficiently show the injurious effects of strong winds, and the necessity of protecting vines as much as possible from their destructive consequences."

But although there can be no doubt as to the evil effects of wind storms, it must be borne in mind that ventilation, and even motion, are essential to the health and growth of the vine. Experiments made by Andrew Knight, show that young trees tied to stakes so as to prevent all motion, do not increase in size as much as those left to the free action of wind. Hence, perhaps, one reason why wire is to be preferred to wood for the cross slats of trellises. In the northern States, however, we in general have wind enough for all useful purposes. But in view of these facts, we would rest content with shelter outside of the vineyard, and unless in very exposed situations we would not deem it advisable to place either trees or fences amongst the vines.

But while we can guard against wind and storms by belts of woods or high fences, there are other influences which we cannot thus alter. Chiefly among these is the exposure of the sun's rays.

Exposure is, in general, derived from one or both of two causes. First, the inclination of the ground, and, secondly, its openness and freedom from overshadowing influences. A wall is a good illustration of the latter—the north side having a northern exposure, and causing fruit planted against it to ripen at a much later period than that planted on the south side, which has a southern exposure. The little raised mounds or flower-beds, to be found in every garden, exhibit the influence exerted by the inclination of the earth—the vegetation on the south side being usually some days earlier than that on the north.

For vineyards, the best exposure is undoubtedly a southern one, slightly inclined toward the east, or at least fully protected from the west, and also from the early morning rays. "It has often been observed that woods or thick trees, buildings, high, broad fences, or steep hills, on the east side of peach orchards, protect the crop. Hence the erroneous opinion, that it is the east winds which do the damage. It is the sunshine upon the frozen buds which destroys them; hence a clouded sky, after a clear frosty night, by preventing sudden thawing, sometimes saves a crop. Covering trees of rare kinds with mats, to shade them from the morning sun, after an intensely frosty night, might sometimes be highly beneficial." (Thomas.)

In this connection, it may be proper to consider

the best direction for the trellises on which the vines are trained. We have often seen a north and south direction advised under the idea that the vines thus receive the sun's rays for a longer time. But the evils attached to this plan are great and insurmountable. In the first place, the vines receive the full force of the early morning sun which, striking the young leaves while still cold, and it may be partially frozen, is productive of the most injurious effects. Then as the day progresses toward noon, the vines are so shaded as not to receive the amount of heat, which they would gladly enjoy at that time, while toward evening again their excitability is greatly increased and is kept up until the last moment, instead of the exciting influence being quietly withdrawn as it ought to be.

But if we give our trellis a direction from east to west, instead of from north to south, the vines will expose but a small surface to the first rays of the sun which will thus warm them gradually, until it attains its meridian splendor, when it will exert its full power and then gradually decline until evening, when everything will gradually cool down. Sudden changes are thus avoided, and the full power of the sun is secured in the ripening of the grapes.

Intimately connected with the foregoing subjects, are the laws which regulate the influence of tempe rature upon vegetation. These are stated by M. De Candolle, as follows:

1. All other things being equal, the power of each plant and of each part of a plant, to resist extremes of temperature is in the inverse ratio of the quantity of water they contain.

2. The power of plants to resist extremes of temperature is directly in proportion to the viscidity of their fluids.

3. The power of plants to resist cold is in the inverse ratio of the rapidity with which their fluids circulate.

4. The liability to freeze, of the fluids contained in plants, is greater in proportion to the size of the cells.

5. The power of plants to resist extremes of temperature is in a direct proportion to the quantity of confined air which the structure of their organs give them the means of retaining in the more delicate parts.

6. The power of plants to resist extremes of temperature is in direct proportion to the capability which the roots possess of absorbing sap less exposed to the external influence of the atmosphere and the sun.

From this it will be obvious that all rank growth and succulent vegetation should be avoided where the desired object is to obtain hardy vines.

Preparation of the Soil

by

John Phin

PREPARATION OF THE SOIL AND FORMATION OF VINE BORDERS.

Having selected a proper site for a vineyard, the next step will be to prepare the soil for the reception of the young vines. It is rarely if ever that ground can be found in a condition fit to plant a vineyard without thorough and extensive improvements, and unless it be in proper order our hopes of success will end in failure and disappointment.

In our remarks on soil it was stated that one absolute necessity is a *dry* subsoil. No other good qualities can compensate for the want of this, and in most cases it is only to be obtained by thorough *draining*.

The first great evil obviated by thorough draining is the existence of stagnant water beneath the surface. It is a saying amongst vine-dressers that "the vine cannot bear wet feet." And nothing can be more true. If the roots be exposed to stagnant water they will become diseased and die off, thus giving rise to weak and ill-ripened though sometimes succulent growth, and hence causing the vine to suffer from the attacks of disease and insects. The grapes, too, will not ripen well, but will remain sour and ill-flavored.

M. Gasparin gives the following observations with regard to the influence which a dry or a moist soil exerts upon the grape : " Other things being equal, we obtain grapes which contain much sugar and little acid from vines grown in a dry soil ; more free acid in a moist soil, and much acid, albumen and mucilage with little sugar in a soil which is absolutely wet."

Another advantage consists in the fact that well-drained land always possesses a higher temperature than that which is wet. This difference amounts to 10° to 12° Fah. and is accounted for by the rapid absorption of heat by the water as it becomes converted into vapor. During this process, too, it is probable that the nascent vapor robs the earth of a portion of the ammonia and gases which it would have separated from the water and retained if it had acted as a filter and the water had passed off by the drains. But however this may be, its effect on temperature is such that Johnson regards thorough draining as equal to a change of climate.

But not only does draining enable the soil to filter all the water which descends upon it, retaining its ammonia, gases and even salts; it is probable that by these means the excrementitious matters discharged by plants, as well as other noxious bodies are washed out of the subsoil or decomposed by contact with the air which penetrates along with the water. In the case of oxide of iron it is probable that a very beneficial effect results from its conversion from the protoxide to the peroxide by means of this influence.

But a change in the chemical constitution and action of the soil is not the only effect of this operation; a no less marked alteration is produced in its mechanical character—heavy lands being rendered light, porous and permeable to the roots of tender plants.

It is unnecessary here to give minute directions for performing such a well-known operation, so we shall merely refer our readers to some of the numerous treatises on that subject. An excellent article on the theory and practice of draining will be found in the "Rural Annual" for 1859 published at the office of the "Genesee Farmer," Rochester, N. Y.

We may state, however, that in laying drains for a vineyard, it should be borne in mind that after the vines are planted it will be almost impossible to get at the drains in case of accident, without serious detriment to the plants. It will, therefore, be well to construct them in the most substantial manner and also to arrange them so that they will not lie immediately under any of the rows of vines. If they are *between* the rows it will not be so difficult to get at them as if they lay directly beneath the plants.

The next great requisite in a soil for the culture of the vine is depth. Ordinary soils of from eight to ten inches are by no means deep enough. Twenty inches is the least depth to be relied upon, and, if very favorable results are desired, it should be made three feet. The subsoil to this depth should be thoroughly loosened, and, unless its quality is very inferior, it may be well to mix it with the surface soil—adding at the same time a good supply of manure or compost. We are aware that some horticulturists object to bringing up the subsoil, but we incline to the belief that if it is of such a character as to produce much injury, the site is unfit for a vineyard. When the subsoil is light (except it be pure sand) no harm can result. If it be pure sand, however, it had better remain where it is unless a sufficiency of clay can be found to mix with it. If, on the other hand, it be so clayey as to hermetically seal up the vine borders, we should prefer to let it remain under. But, if possible, a site should be selected where a good depth of tolerable soil may be obtained either naturally or by proper effort.

The advantages incident to depth in ordinary cases consist in the roots being placed alike beyond the extreme heat of summer and the severe cold of winter. Consequently they do not suffer from drought, and are able at once to enter upon their duties in the spring.

For table grapes, we doubt whether the soil can be too deep or rich—not meaning by the latter term, however, saturated with *undecomposed* organic matter. But observation leads us to doubt the propriety of carrying these features to an extreme in the case of closely-trimmed vines cultivated for wine. It is true that the Western authors (Remelin, Buchannan, etc.—some of them Europeans) advocate this depth and richness. But, if our memory does not deceive us, some of Mr. Longworth's tenants who have not pursued the most thorough system of cultivation have occasionally escaped evils to which their more *skillful* and hard-working brethren have been exposed. And perhaps a solution of this mystery may be found above, notwithstanding Mr. Longworth naïvely tells us that he cannot believe that nature ever favors the indolent. Our own experience in this particular department is not sufficient to warrant us in pronouncing a decided opinion on the subject; but the principles of physiology would lead us to believe that if the roots of vines are planted in a deep and rich soil the branches must be allowed corresponding elbow room. If we desire to keep a vigorous plant down we must starve and curtail its roots as well as use the pruning-knife on its branches.

There are two methods of deepening a soil, viz: by the subsoil plough and by trenching with the spade. Both these operations are too well known to require a minute description, though in regard to the latter there are so many and such contradictory directions given in books that we may be pardoned a few remarks in relation thereto.

In order properly to trench a piece of ground the directions given by Loudon are as explicit and judicious as possible. "Trenching is a mode of pulverizing and mixing the soil, or of pulverizing and changing its surface to a greater depth than can be done by the spade alone. For trenching with a view to pulverizing and changing the surface, a trench is formed like the furrow in digging, but two or more times wider and deeper; the plot or piece to be trenched is next marked off with the line into parallel strips of this width; and beginning at one of these, the operator digs or picks the surface stratum, and throws it in the bottom of the trench. Having completed with the shovel the removal of the surface stratum, a second, third or fourth, according to the depth of the soil and other circumstances, is removed in the same way; and thus, when the operation is completed, the position of the different strata is exactly the reverse of what they were before. In trenching with a view to mixture and pulverization, all that is necessary is to open, at one corner of the plot, a trench or excavation of the desired depth, 3 or 4 feet broad, and 6 or 8 feet long. Then proceed to fill the excavation from one end by working out a similar one. In this way proceed across the piece to be trenched, and then return, and so on in parallel courses to the end of the plot, observing that the face or position of the moved soil in the trench must always be that of a slope, in order that whatever is thrown there may be mixed and not deposited in regular layers as in the other case. To effect this most completely, the operator should always stand in the bottom of the trench, and first picking down and mixing the materials, from the solid side, should next take them up with the shovel, and throw them on the slope or face of the moved soil, keeping a distinct space of two or three feet between them. For want of attention to this, in trenching new soils for gardens and plantations, it may be truly said that half the benefit derivable from the operation is lost."

A more expeditious method of mixing the soil, and one which varies but slightly from the ordinary system, consists in cutting down the bank in successive sections so as to produce theoretically a series of layers of soil and subsoil, but in reality a most inti-

PREPARATION OF THE SOIL.

mate mixture of the two. This is best accomplished by opening a very wide trench—say from four to six feet wide. Then throw the top spit off a bank of the same width into the bottom of the trench so as to insure the burial of all insects, seeds, and weeds; cut a width of from six to fifteen inches of the remaining portion of the bank completely down to the bottom, and spread the soil so obtained in a thin layer over the spit formerly thrown in. Then cut down another six to fifteen inches in the same manner, proceeding thus until the whole bank has been cut down and used to fill up the trench. It will now be found that, with the exception of the extreme top spit which is placed at the bottom for very good reasons, the whole soil is sufficiently mixed for all practical purposes.

Another mode of trenching—called bastard trenching—is thus described by a writer in the "Gardener's Chronicle:" "Open a trench two feet and a half, or a yard wide, one full spit and the shovelling deep, and wheel the soil from it to where it is intended to finish the piece; then put in the dung and dig it in with the bottom spit in the trench; then fill up this trench with the top spit, etc., of the second, treating it in like manner, and so on. The advantages of this plan of working the soil are, the good soil is retained at the top—an important consideration where the soil is poor or bad; the bottom soil is enriched and

3*

loosened for the penetration and nourishment of the roots, and allowing them to descend deeper, they are not so liable to suffer from drought in summer; strong soil is rendered capable of absorbing more moisture, and yet remains drier at the surface by the water passing down more rapidly to the subsoil, and it insures a more thorough shifting of the soil."

A method which we have sometimes adopted, and which we think a saving of labor under some circumstances, is as follows:

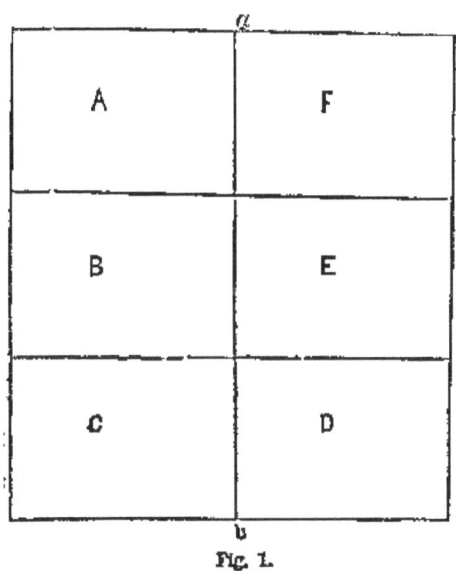

Fig. 1.

Let fig. 1. represent the plot of ground to be trenched. Divide it into two equal parts by the line *a b*, and instead of wheeling the soil out of A F to the rear of the plot, simply throw that from A out in front.

PREPARATION OF THE SOIL.

There can, of course, be no more difficulty in finding room for it there than there would be in obtaining a place for it in the rear. Then dig down the bank B, and with it fill the trench A. B is now a trench which may be filled from C; C may be filled from D; D from E; E from F; and the filling of F with the soil which was at first thrown out of A, will make all even. The wheeling of the soil, which is no inconsiderable item, is thus saved. It is evident, however, that this plan is adapted only to small, or at least narrow plots.

All the foregoing operations prove most beneficial when performed in the fall. At that time the soil should not be finely pulverized, but left in as rough a state as possible so as to expose it thoroughly to the action of the winter's frost and snow. It should be also well mixed with a good dressing of well decomposed stable manure, and any of those matters mentioned in Chapter XI.

By these means, the ground will be thoroughly enriched by spring, and will not consist of earth mixed with fermenting masses of manure, than which nothing can be more injurious to young plants. In the following spring the land should be raked or harrowed, so as to obtain a level surface of finely pulverized soil, and if it should be lightly forked over it would be none the worse for it.

TERRACES.—From our directions for the selection of a vineyard site, it will be seen that we prefer a gentle slope to the south or southeast. If this slope does not exceed an angle of eight degrees, or a rise of one foot in seven, it will be unnecessary to adopt any peculiar system of arrangement. For a rise of one in four it will be necessary merely to make very slight terraces, the borders being made eight feet wide and half the descent being taken up by the slope given to them, will leave but twelve inches of a terrace, which may be easily secured by a row of sods, boards or stones, or even the earth beaten hard and kept carefully dressed up. But when the inclination of the ground much exceeds this amount, it becomes necessary to form regular terraces which is best done as follows:

Find out the actual slope or inclination of the ground, which is easily done by taking an eight-foot

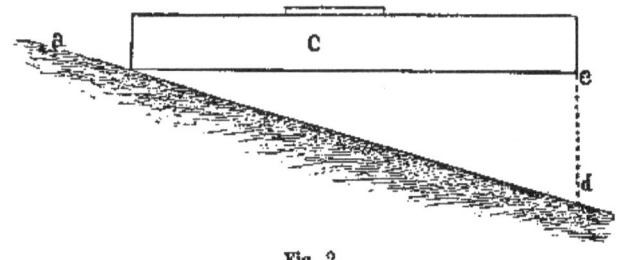

Fig. 2.

board, and after laying one edge on the ground and levelling the board, find the length of the perpendicular which touches the surface beneath the other end. Thus *a d*, fig. 2, being the surface of the hill, and c the eight feet board with the level resting upon it, *e d*, will be the rise in eight feet and *e d*, less the slope given to the border will be the height of each step or terrace. Having found this, the next step is to cut a perpendicular face half the height of the proposed terrace at the foot of the hill and against it to build a wall as high as may be required. This is best formed of dry stone, though the bank is sometimes left with a good deal of slope, and sodded, the sods being pinned to the face of the bank with stakes until the roots have penetrated sufficiently to hold. The sods for this purpose should not be cut square, but diamond form, so that the face of the bank would present the appearance shown in fig. 3. But sods are

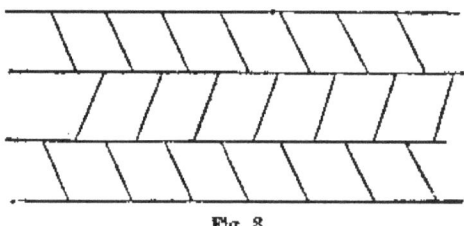

Fig. 3.

objectionable from the fact that they not only keep the air moist in the vicinity of the vines, but also abstract a good deal of nutriment from the soil, and unless kept neatly mown present a very bad appearance. In default of good stone we think that sun-dried brick would make a very good wall. The earth of which they are made should be mixed with straw, well worked and made into blocks.

It is probable that in well-drained terraces such walls would last well if protected with a coping of boards or straw secured with good clay in the manner shown in fig. 4, so as to shed the rain.

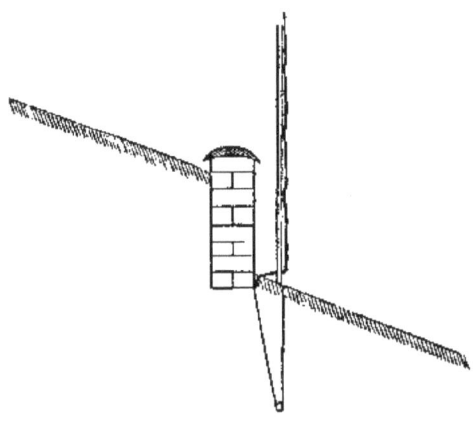

Figure 4.

Having built this wall, the next step is to fill up behind it, and level off a border of suitable width— say 6 or 8 feet. To do this it will be necessary to cut down a perpendicular face the same height as before, when another wall must be built, and the same process repeated.

A writer in the third volume of the "Gardener's Magazine" proposes to train the vines on trellises lying on the surface of the slopes as shown in figure 5. Trained in this manner, grapes are said to have

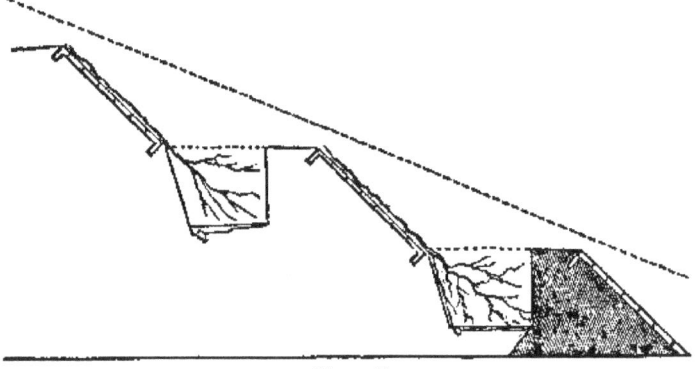

Figure 5.

ripened well in England. We would prefer the vertical trellis, however, and give the illustration, more to show what has been proposed than as an example to be followed. So many times have we seen it proposed to incline trellises and train vines horizontally, that we cannot refrain from quoting Lindley's remarks upon this point.

"That training a tree over the face of a wall will protect the blossoms from cold must be apparent, when we consider the severe effect of excessive evaporation upon the tender parts. A merely low temperature will produce but little comparative injury in a still air, because the more essential parts of the flower are very much guarded by the bracts, calyx and petals, which overlie them, and, moreover, because radiation will be intercepted by the branches themselves, placed one above the other, so that none but the uppermost branches which radiate into space will feel its full effects; but when a cold wind is constantly passing through the branches and among the flowers, the perspiration—against which no sufficient guard is provided by nature—becomes so rapid as to increase the amount of cold considerably, besides abstracting more aqueous matter than a plant can safely part with. To prevent this being one of the great objects of training trees, it is inconceivable how any one should have recommended such devices as those mentioned in the 'Horticultural Transactions,' II. Appendix, p. 8., of training trees upon a horizontal plane; the only effect of which would be to expose a tree as much as possible to the effect of that radiation which it is the very purpose of training to guard against."

All terraces should be well drained, and the drains are best arranged by having a series of cross drains parallel to the terrace, as seen in section fig. 4 and 5, and emptying into a main drain which descends the hill. These drains should be placed as in the figures, taking care to leave the ground under the wall solid and undisturbed. In forming terraces for vine culture it is necessary to exercise care and judgment, so as not to bury the good soil and leave the poor soil for the vines to grow in.

FORMATION OF VINE BORDERS.

VINE BORDERS.—The formation of vine borders in gardens is a subject upon which the student will find no lack of information, almost every successful gardener attributing the superiority of his grapes to some peculiarity in the construction of his borders, and innumerable have been the paper conflicts waged between the advocates of carrion, asphalte, ventilated borders, etc., etc., and their opponents. The "carrion" controversy has probably caused the shedding of more ink than any of the others, the ultraists on both sides being probably in the wrong. But, after all, we regard the construction of proper vine borders as no very difficult affair, and shall first give our own views in the matter and afterward quote those of other authors.

Of course in borders, as in other cases, it is necessary that the bottom be as dry as possible. This being provided for, if the soil is a light mellow garden mold, we would rest content with trenching it thoroughly, and adding liberal supplies of litter, well decomposed manure, woollen rags, and especially bones;* and if in the bottom of each trench a good

---

* In the ordinary course of agriculture, where "quick returns," if not "small profits" are an important element of success, bones when used as manure cannot be too thoroughly pulverized. Indeed, it is often profitable to reduce them to the most active form—that of a solution—by means of acids. But for reasons to be hereafter stated, one

layer of brickbats, lime rubbish, and oyster shells be laid, it will prove an advantage. A border prepared in this simple manner will give good satisfaction under any circumstances.

If the soil be heavy we would also make liberal additions of sandy loam or saw-dust.

But if the location of the border is such that it cannot be well drained, we would remove all the soil to the depth of 18 inches over the entire extent of the border and fill up at least 12 inches of the space with stones, brickbats, etc. Over this we would spread a thin layer of straw or brush, and after building a wall round the border 18 to 30 inches high, we would fill in with a rich soil resembling in composition, that described above. The earth on the outside might be banked up to the wall, and either sodded, or merely beaten solid.

In all such cases, it is evident that from the narrow limits to which we are in general confined, the soil ought to be of the richest kind; and as it is nearly impossible to renew it after the vines are once started, this richness should be derived from materials calcu-

---

great advantage to be derived from the use of bones in vine borders is the length of time during which they continue to act, and, therefore, the largest and most solid should be selected and used without being crushed or broken. This is no argument, however, against the *additional* use of bone dust.

lated to give more than a mere temporary impetus to the plants. The nature and action of manures will form the subject of a future chapter, but we may here state that bones, hair, woollen rags, leather clippings and similar matters are by far the most suitable. For the purpose of giving porosity to the soil, as well as furnishing nutriment to the plants, nothing will be found to equal chopped straw. Chaff, or sawdust comes next in order, and from experiments which we have made on the subject, we do not think the value of the latter is half appreciated. To dead animals, either whole or divided, we have never found any objection, provided they were not placed in direct contact with the roots of the plant. No danger is to be apprehended of the vine seeking them to its own detriment. But this more properly pertains to the subject of manures. We will now give the manner in which the most celebrated grape growers construct their borders.

Miller (1759) recommends good mellow soil without any addition.

Speechly (1790) states in his work: "As the vines in the hot-house at Welbeck have been remarkably fruitful and vigorous, I shall beg leave to recommend the same kind of compost mold which I make use of there, viz. one-fourth part of garden mold, (a strong loam); one-fourth of the sward or turf from a pasture where the soil is a sandy loam; one-fourth of the sweepings and scrapings of pavements and hard roads; one-eighth of rotten cow and stable yard dung mixed; and one-eighth of vegetable mold from reduced and decayed oak leaves. These are the several and respective proportions. The sward should be laid in a heap till the grass roots are in a state of decay, and then turned over and broken with a spade; let it then be put to the other materials, and the whole worked together, till the separate parts become well and uniformly mixed and incorporated.

As the vegetable mold from decayed leaves cannot always be obtained, by reason that the leaves require two years before they become sufficiently putrid and reduced, it therefore may sometimes be necessary to substitute some other ingredient in lieu of this part of the compost; wherefore it may not be inexpedient to point out the proper succedanea.

Rotten wood reduced to a fine mold, such as is often found under fagot stacks; the scraping of the ground in old woods, where the trees grow thick together; mold out of hollow trees, and sawdust reduced to a fine mold, provided it be not from wood of a resinous kind, are in part of a similar nature with vegetable mold from decayed leaves, but are neither so rich nor so powerful, because the vegetable mold receives a power by its fermentation.

FORMATION OF VINE BORDERS.

Abercrombie directs the top slip of sandy loam from an upland pasture, one-third part; unexhausted brown loam from a garden, one-fourth part; scrapings of roads free from clay, one-sixth part; vegetable mold or old tan, or rotten stable dung, one-eighth part; shell marl, or mild lime, one-twelfth part. His borders he recommends to be from three to five feet in depth, and where practicable, not less than four feet wide within the house, and not less than ten feet wide without.

The vine borders at Wishaw House, Lanarkshire, in a cold and wet locality, are thus formed: Breadth, 12 feet, depth of soil 18 inches, under which is laid a foot of hard clinkers, by way of drainage. The soil used is that natural to the garden, which had for years been under pasture, and is a remarkably strong, rich brick-clayey loam, with no other preparation than the addition of a moderate supply of stable manure. In this soil the best grapes ever produced in Scotland have been grown for the last three years.

A writer in the "Gardener's Chronicle" (1843, page 825) prepares his borders thus: The soil most suitable for a vine border is the surface spit from a field of an old fertile loam pasture; this should be collected some time before it is required, mixed with a good proportion of cow dung, and the whole turned over at intervals, three or four times, and exposed to the action of the weather. In preparing the border, the old earth should be cleared away from the whole space, to the depth of about two and a half feet, and a main drain cut parallel with the length of the border, at its extreme outer edge.

This should be at least two feet lower than the bottom of the border, whether laid with concrete, chalk or bricks, and the bottom of the border should have a gentle inclination from the back to the drain. To render this drainage more effectual, cut small drains, placing drain tiles at their bottoms, at convenient distances, to run in a slanting direction from the back of the border into the main drain, the latter being six inches below them. A few turfs should be laid over the tile drains with the grassy side down; the fresh soil may then be filled in, taking care to keep the roughest part near the bottom.

Three cubic yards of compost are enough for each vine; this will admit of the border being ten feet wide, or with forty-eight cubic feet, you may form it only six feet wide in the first instance, and add six feet more as the vines extend.

Roberts, the great advocate for carrion, gives the following description of his border: "The compost and manures I most recommend, and which I made use of, are two parts the parings of a piece of old pasture land, a strong loam laid up one year (or till

FORMATION OF VINE BORDERS.

the sward is half decomposed), in the form of a potato hod, close covered in with soil, and never turned; one part, the turf with four inches of the soil, of a looser texture laid up for the same period, and not turned, as before; an eighth part scrapings of the highways formed from limestone, or other hard material; and the other eighth part, half decomposed horse or cow dung. I am not an advocate for turning over and mixing the materials promiscuously together, as, by often turning, the compost becomes too solid, losing a great portion of its fertilizing property by such repeated intermixture; and unless it be of a very sandy, loose texture, the border will, in a few years, become impervious both to water and to atmospheric air, which are of incalculable benefit to the growth of the vine. I would recommend the autumn, if the weather be dry, to prepare to fill in your border.

"A month previous to filling your border, provide a quantity of carrion, cattle dying by accident, disease, etc., which I am sorry to say, has, of late years, been too common an occurrence. If you have collected it some time before hand, have it cut into small pieces and laid up in soil till the time of using. It emits a very nauseous effluvia, but this must be borne, for this is the *pabulum* to produce the nectar of Bacchus. When all is ready, and the weather favorable, proceed at one end of your border, wheeling in and mixing the materials in proportion as they stand to each other in my previous directions, on no account breaking the materials in mixing, but turn them in as rough as possible, adding one good sized horse or cow carcass to every ten or twelve square yards, using caution, and not bringing it to the surface of the border within one foot as its assistance is not wanted the first year. What I have here recommended is my practice adopted at this place, the result of which, I dare presume to say, has surprised all, both gentlemen and practical gardeners, who have witnessed it."

Fiske Allen, one of the best American culturists of the vine under glass, constructs his borders thus :

" If the soil is very poor, or unsuitable for the purpose, so as to require to be removed entirely, then a compost prepared thus is recommended; one-half to be the top soil of an old pasture, one-quarter to be bone, or some other strong manure; one-eighth oyster shells, or lime and brick rubbish; one-eighth rotten manure; these articles thrown together in a heap, and so to remain until decomposed and amalgamated, when they should be placed in the border and thrown loosely together. My borders having the most slaughter-house manure, or whole bones of animals in their composition still continue, as they ever have done, to produce the best fruit and the largest crops.

FORMATION OF VINE BORDERS.

"It is unnecessary to attempt to give rules for every kind of soil. One must use his own judgment, and make his border to consist, as near as can be, of the above ingredients. He must bear in mind that, if his soil is a stiff clayey loam, he must add freely of such materials as will lighten and give permeability to it. If the soil is light, sandy or gravelly, with the manure should be added a proportion of clay or clayey loam. The rich alluvion soil, abounding in our western and southwestern States, will not require any of these strong manures. If anything is requisite to improve them, it must be shells, charcoal, leaves, small stones, or gravel—such materials as will loosen the soil."

But that the reader may not be discouraged by these extravagant demands we quote the following from Hoare:

"But if vines could not be planted with any prospect of success in any other situations than in borders set apart for that purpose, but a very small quantity of grapes could be grown, compared with what the country is capable of producing. Innumerable instances occur throughout the country, and especially in towns and their suburban districts, in which walls, cottages, houses, and various descriptions of brick and stone erections present very favorable aspects for the training of vines, but which nevertheless are so situated locally, as to possess little or no soil at all on the surface adjoining their sites; the ground being either paved with bricks or stone, or perhaps trodden so hard, as to be apparently incapable of yielding sustenance to any vegetable production.

"In all such cases, however, if the ground adjoining the site of the wall or building be opened to the extent of eighteen inches square, and as many deep, it will be sufficient to admit the roots of a young vine, which must be pruned to suit that space. If a wider and deeper space can be made, it will of course be better; but if not, that will do. After the sides and bottom have been loosened as much as possible, the vine may be planted and the hole filled up with two-thirds of rich loamy earth, and one-third of road scrapings, previously mixed well together, and if necessary the surface covering, whether of stone, brick, or otherwise, may be restored again to its former state, provided a space about six inches square be left open for the stem to swell in during its future growth. Vines planted in such situations, will in general do well, although their growth will not be so rapid as when planted under more favorable circumstances.

"In all cases where vines are planted against any description of buildings, their roots push as soon as possible under the foundations, being attracted thither by the warm air which is there generated; and such situations being also dry, from the excavations which have been made, offer to the roots the same protection from excessive moisture, as the substratum of a well-prepared border. The same may be observed of vines planted against walls, the foundations of which possess similar advantages, although in a more limited degree. Hence the fact may be inferred that vines planted in such situations, without any previous preparation of the soil, will frequently grow as luxuriantly, and produce as fine grapes as those planted in rich and well-prepared borders.

"Indeed, it is hardly possible to plant a vine in any situation in which it will not thrive, provided its roots can by any means push themselves into a dry place, and the aspect be such as to afford to its branches a sufficient portion of the sun's rays to elaborate the juices of the plant.

"The truth is, that the roots of the vine possess an extraordinary power of adapting themselves to any situation in which they may be planted, provided it be a dry one.

"They will ramble in every direction in search of food, and extract nourishment from sources apparently the most barren. In short, they are the best caterers that can possibly be imagined, for they will grow, and even thrive luxuriantly, where almost every description of plant or tree would inevitably starve."

Soil and Cultivation

by

William C. McCollom

Soil and Cultivation

Soil — Manure vs. fertilizer — Necessity of trenching — Building up run-down arbours — Mulching — Cultivation of the soil — When and how to water.

SOIL is a question of utmost importance in the growing of vines; most of the vine family are rank growers and demand a liberal quantity of plant food. If you give them what they require, you are almost sure to have good results. If you plant them in poor soil, without any preparation, do not look for anything but failure. I should estimate that ninety per cent. of the disease and trouble we encounter in growing vines arises from improper preparation of the soil, or none at all.

For vines of all kinds, I prefer manure to commercial fertilizers. It is more lasting, it collects and stores so much moisture that it encourages root action, and, finally, there is very little danger of giving the plants too much and thereby injuring them. For very heavy soils it may be advisable to cut down on the manure and use some fertilizer, but, all things considered, manure is far ahead of any fertilizer we have to date. The best way to apply the manure, so that the plants will get the most benefit from it, is by trenching or subsoiling. That is, by making the top and bottom soil change places — a very simple operation if you get started right.

For a pergola or arbour, mark off a bed six feet out from the posts. Go to the end of this bed and measure off four feet, remove the soil to the depth of three feet, then put about six inches of manure in the bottom of the trench. Mark off four feet again from the edge of the trench, throw about one foot of top soil into the bottom of the previously made trench, then about six inches of manure, and another layer of soil, repeating until you have made three layers of soil, when the trench is finished. For real heavy soils I would recommend placing a little drainage in the bottom of the trench — some old bricks, clam shells, or anything that will not retain water. This may seem a lot of expense to go to, but I am sure that any one who tries it will never regret it. The best cheap substitute is a good deep ploughing, turning under a liberal quantity of manure, and having a subsoil plow follow in the rows to loosen up the bottom as much as possible. Vines, more than other plants, require this deep trenching because their roots, instead of growing out horizontally, as those of many trees do, have a tendency to grow downward.

I once had occasion to transplant a wistaria which had been growing in the same place for some fifteen years. When I dug down around the roots I found one big root running straight down. I dug down about six feet and was then obliged to cut it, as the end was not yet in sight.

In planting a vine at the base of a tree or some place where only a single specimen is required, make a good, deep hole, the same as you would for a tree, and put some manure in the bottom. Make it a miniature trench and you will accomplish in two seasons what it would take ten to achieve after merely sticking the plants in the ground.

For old, run-down arbours, or where vines on a building are beginning to deteriorate, go out a safe distance from the plants and trench the soil as you would for a new planting; or, instead of letting the vines run down, give them a semi-annual application of some good fertilizer. Mulching is very beneficial to the plants, and, where possible, is certainly a good practice.

SOIL AND CULTIVATION

Put on a good mulch of manure in the fall, and the winter rains will wash a great deal of the nourishing qualities into the soil. In spring the manure can be turned under. This, too, will prevent your vines from getting run down too quickly.

Cultivate the soil deeply and as often as possible, if you desire your vines to grow well. This is very important; it should be done, at least, after every rain, although, of course, the oftener the better. Vines are deep rooters, therefore deep cultivation is preferable. Loosen up the top soil well with a digging fork, using a little care; do not go too close to the stem of the plants, as there you may find a few surface roots.

Water is a very important factor in the growing of vines, and the first question that arises is, when is the proper time to apply water? In good, rainy seasons, it probably will not be necessary to apply any, but we all know that nine out of ten seasons, plants suffer a great deal from drought in this country.

The common error made in the watering of vines is simply this: the vines receive no water until they are ready to flower. When the flower spike appears, the hose is hauled out of the cellar and watering started. Plants that flower but once a year require water when in growth, as they are then producing the wood. They require less water when in flower than at any time during the year. For example, a spring-flowering vine will require plenty of water immediately after the flowers have gone, when growth commences, while a fall-flowering vine will require water as soon as growth starts in the spring. Thus the important watering periods are spring and early summer, and, if the weather is dry during May, June, and July, get out the hose and give the plants a thorough drenching occasionally. Do not merely give the ground a sprinkling, but let the hose run on the border for a whole day; then it will not be necessary to repeat the operation so often. The lighter the soil the oftener it must be watered. If the soil is properly watered, you should be able to ram a rake handle well down into the border at any time.

Location and Soil, Preparation of the Ground and How to Cultivate the Soil

by

Charles Reemelin

LOCATION AND SOIL.

THE best location, if it is desired to produce good wine, is the southern exposure of a hill or hillock. There the vines get the proper sunshine, and are also properly protected against storms, especially the north and north-east winds. A location giving an exposure midway between east and south is also favorable, because such an exposure gets the sunshine from morning till pretty late in the evening.

LOCATION AND SOIL.

A due eastern exposure is less favorable, since it loses the sun too early; it is ever exposed to eastern winds, and is sensitive to frosts, even of the lighter sorts, because it receives the rays of the sun so very early and direct, as to subject such locations to injury from freezing nearly every year.

Still worse is a western exposure, because it receives the sun till very late, and hence suffers from the chilly evening dews, which in this country are, comparatively speaking, far colder than in Europe. Such an exposure must necessarily suffer from west winds, and is also more liable to be injured by hail.

Hills and hillocks are far better for wine-culture than plains, which latter may produce greater quantities of wine, but it is invariably of a poorer quality. Plains or hills whose soil, either on the surface or as a substratum, has yellow or blue clay soil, are not favorable for vineyards, because upon such soils neither the atmosphere, nor the sun, rain or dew, can operate as they should, and hence there is danger that the vine will be affected with the wet-rot. The sun's rays hardly ever strike vines upon plains in the proper direction, so as to afford the required warmth, and the wood and the grapes are apt not to get the proper ripening; such localities are also far more subject to winter and spring frosts, and to mildew. And, in addition, they must necessarily suffer more from destructive insects, snails,

1*

and animals of every description, as it is well known that such prevail more largely upon plains.

There exists, however, a great difference between the foot, the middle, and the crest of hill-sides. The middle gives the best wine; the foot is more or less subject to frost, and does not receive adequately the sun's rays; while the crest is too much exposed to cold winds, in addition to its soil being very seldom good enough.

Nor must neighboring objects be lost sight of in locating vineyards. Favorable is everything which tends to temper and somewhat increase warmth, and which protects against frosts, raw winds, and other casualties; such as woods, buildings, high walls, and adjoining hills, provided they are in the rear or north of vineyards, and such localities will always produce the earliest ripe fruit and the best quality of wine.

Injurious objects, when too near neighbors, such as lakes, ponds, swamps, and cold wet woods, are to be avoided, as from all these cold mists are apt to generate. Hills, houses, trees, &c., should not be so near as to throw a shade over the vineyard. Vineyards should never be planted along deep valleys, hollows or gorges, which run east and west, since such almost invariably produce, in winter especially, constant drafts of wind, and they are more or less injurious. Grass and clover patches should not be too near, as

LOCATION AND SOIL.

they draw frosts, and smithies or other large laboratories or manufactories, are also to be avoided on account of the smoke.

Much depends, also, upon the quality of the soil, which changes often within a very small space of ground. A grapevine will grow, to be sure, where other plants grow, but the quality of the wine is always modified by the kind of soil.

A heavy soil—one composed of sticky clay—will not permit sun and rain to penetrate, and may therefore be termed a cold soil. In such, grapevines soon become weak and sickly; in wet seasons their foliage is apt to have a yellowish tint; the roots rot, and even where that should accidentally not take place, the quality of the wine will never be very good. The only way to render such a soil fit for a vineyard, is by a copious application of lime or marl, mixed with sand,—yet it may be done, but not efficiently, by mixing with it a sandy loam. Little, however, as vineyards will prosper in such cold soils, they will succeed just as little in too light sandy soils, unless well mixed with clay loam, or clayey marl.

The soil most to be preferred in climates such as the Northern and Middle States of this Union, is that so generally prevailing rich loam, mixed with some gravel and marl. This kind of soil differs largely in various locations, and it will take a more or less lengthened period of individual experience to find the best locations. We should, however, always examine into the more general admixtures of the soil which we propose to select for our vineyard. There should always be some sand, some clay, some limestone, and some gravel in it. Is there too little sand or gravelly limestone, then, the soil will soon become too clayey and cold, or if there is too much gravel and sand, then vegetation is impeded. Our soils almost invariably lack what European vineyard men prize so highly—gypsum—and this must be supplied by proper manures. The best wines in Europe grow upon the hill-sides of lime or gypsum mountain ridges, whose formation is somewhat mixed with sandstone. The color of the soil is not always a sure indication of the quality of the soil, but it may be assumed as a general rule, that soils for vineyards should neither be a very light yellow, nor a very brown red. To recapitulate, therefore :

The altitude of a vineyard should neither be too high nor too low, as compared with the surrounding country. The exposure should be selected with due reference to giving full chance to the sun's rays during the entire day; and the soil should neither be too rich nor too poor,—affording to the roots of the vine and to atmospheric influences an easy chance to penetrate.

PREPARATION OF THE GROUND.

This matter embraces the foundation of the whole subject of vine culture, and herein nearly every vineyard yet planted in America is defective. Labor is so extremely high here, as to make it seem to us almost impossible to start a vineyard as it should be. Our very best vineyards are spaded up but two feet, while in many parts of Europe they spade up the ground to the depth of three and four, and even five feet. We never prepare the ground itself, during the preceding year, while in Europe it is sowed down in clover, for a few years previous, and well covered with good coatings of gypsum and manure. We trust to the virgin richness of our soil, and in our confidence are apt to forget that spading up the ground for several feet is done for other reasons besides mere fertilizing; and that among these, for us especially, must be a sinking below the subsoil the present surface or upper soil, which being full of decomposed vegetable matter, is the hot-bed of all manner of insects. The spading up and turning of the surface soil beneath its present subsoil, is of the greatest importance, because thereby the "Foot Roots" may penetrate downwardly, and thus give to the whole grapevine not only its vigor, but also its great safeguard against too sudden atmospheric changes, or long-continued droughts. And I may in connection with this, here remark upon an erroneous suggestion, which I have noticed in some agricultural journals. They suggest a longer "stem." I do not think that the stem should be much longer than twenty inches, but think it of the first importance that the foot roots should penetrate deeply.

The ground intended for a vineyard should be well manured the previous year, either by a coating of lime, where that kind of manuring is proper, or by gypsum, where it can be had; or by ploughing under some green sward, such as clover; or at least by a good and thorough coat of manure, straw, or even leaves.

Of the ground thus prepared, the surface should, for the depth of twelve inches at least, be sunk beneath twelve inches of soil immediately underneath. This is best done, if the ground be loamy, with the spade, or if stony with the mattock. For this purpose a trench is first dug four feet wide, and to the depth to which the vine-dresser is going to spade up and trench his vineyard. Into this first trench, say four feet wide and two to four feet deep, and as long as the vineyard may be, say two hundred feet, is then thrown twelve inches of the surface soil (using the very best steel spades), and by driving the spade into the ground as nearly perpendicular as possible, and not slanting, as lazy laborers are apt to do; for thus alone can this top soil be spaded up to the depth of at least twelve inches. The loose soil

PREPARATION OF THE GROUND.

which is left in the trench, having crumbled from the spade, must then be carefully scraped into the first trench, and then the twelve inches of subsoil must again be similarly spaded up and thrown upon the previously spaded up surface soil, and so on, each twelve inches to the depth required. And the loose soil left in the bottom, must also again be carefully shoveled up and thrown upon the other ground. Thus trench after trench will be regularly formed, until the whole allotted piece is finished. Let the reader bear in mind, as the *sine quâ non* of a good vineyard, that it is not a mixture of the surface with the subsoil that's wanted; but that the subsoil cover, for twelve inches at least, and twenty to thirty if possible, the original surface soil, and the deeper this is done (always in reason) the better. It is far better, view it in whatever light we may, to have a small, good vineyard, than a large, poor one.

The ground thus spaded up should be permitted to settle well, before the vines are planted. One or two good rains will generally accomplish this. The best method is, however, to trench in the fall, and plant in the spring.

There are other methods of preparing the ground. One is to make large holes, throwing the surface soil underneath and planting the vines therein.

Deep ploughing and subsoiling is also frequently

adopted. I have tried all these methods. The first vineyard I set out by merely digging holes; another by ploughing some sixteen inches deep, with a large plough, drawn by four yoke of oxen, and followed with a subsoil plough, drawn by a pair of horses, and another by trenching as above suggested, thirty inches deep. As to results I can only say, that the first planted vineyard is now being dug up, because it was always liable to every disease which happened to prevail in the season, having hardly yielded a fair compensation for the labor expended; the subsoiled vineyard does better, but I have no hopes of its lasting more than twenty years; while a well-trenched vineyard, to the depth of thirty-six inches, with such virgin soil as we have in America, should, and doubtless would last—*if otherwise properly managed*—eighty to one hundred years. I shall hereafter trench any vineyards I may plant, at least thirty-six inches, and recommend the same course to all others.

I am informed that there is now being constructed in Cincinnati, a large plough to be drawn by six yoke of oxen, and warranted to plough the ground twenty-eight inches deep. I have not seen this latest improvement, and can only say that unless this plough does leave a clean furrow, at least twelve inches wide of the promised depth, it will not answer. The large ploughs I *have* seen do not accomplish this. They break the ground

PREPARATION OF THE GROUND.

up, mix it somewhat, but do not turn the top soil under. This, for reasons already stated, is not enough.

It is hardly necessary for me to say, that the procedure must be varied with the ground. Some soils are naturally rich to the required depth, though I should fear such soils for vineyards. Others are very rocky and must be worked with the mattock and grubbing hoe. Good sense will in each case dictate the requisite mode, if we will but bear in mind the great point in a vineyard view. This is to get the surface soil beneath the subsoil, so as to afford from the very start of the vine, to the "roots" at the "*foot*" of the vine, an easy, healthy and steady downward growth. They are the life of the vine, and their continued health is most important. If they are but thrifty, then we need not fear but what the "side" and "surface roots" will always grow and prosper in due time and in proper manner.

In vineyards along side hills, it is well to use the stones generally found therein, for the purpose of erecting walls to prevent "washing." These walls should have their foundation deep enough, so as to be out of the reach of heavy winter frosts. They should be so slantingly laid up, as to bear properly "to land." Such walls are not only useful, but they are an ornament to the vineyard and the general landscape. If properly laid up, they last as long as the vineyard.

Where stones are lacking, it may be necessary to

raise banks by sodding them with green sward. They are not as good as stone walls, since the green sward is apt to subject the neighboring vines to frost, but the ground must be protected from washing even at this risk. I take it for granted, however, that there are very few side hills indeed, where by trenching deep enough, there will not be the required quantity of stones.

I have thus indicated the general rules by which we must be guided in the preparation of the ground in each special case, and I must now only add, that it is a great but frequent error to suppose that throwing old logs, brushwood or stones, underneath, promotes the growth of vines. They may not hinder them, if well packed with ground, but great care should be taken not to leave vacuities, as they are sure to impart to the "foot roots" an unhealthy state. Vines should always be planted after the ground is well settled, and not before. The ground should also be well harrowed, so as to render it perfectly even and in complete cultivating order.

Before dismissing this chapter I would add, that according to my experience, there is, in fact, but little actual difference in the *cost* between a well-trenched vineyard and one slovenly laid out. To trench an acre three feet deep, is worth in common soil $100; two feet deep, $75. With large ploughs, followed by subsoilers, an acre costs about $25. To dig holes, merely costs about $15. But mark it, you save in a well-trenched

PREPARATION OF THE GROUND.

vineyard each year, for three years, *one* hoeing, at least, and you get a good crop in the fourth year. Your vines grow up regular, as in such a vineyard but few miss; and lastly, let me say to you, that having started right, you are apt to keep right, and are therefore every way sure to have a good vineyard.

HOW TO MARK OUT A VINEYARD, AND GET IT READY FOR THE VINES.

The ground being properly prepared and settled, as previously suggested, the next thing to be done is the staking out of the vineyard. To do this, it is well to prepare as many little marking sticks (say twelve inches long and half an inch square, pointed at one end) as there are to be grapehills in the vineyard; for instance, an acre planted four feet apart each way, about 2,200 hills.

The next thing to determine is, how far apart it is intended to have the rows. There is no settled rule upon this subject. I have seen, in Europe, thrifty vineyards one foot apart, and I have seen them ten feet apart. The four by four may, however, be said to be the prevailing and most approved custom. I have myself lately adopted five by five, and I like the appearance of it very well. In Italy, I am told, vines are planted

HOW TO CULTIVATE THE SOIL.

Never put a "Spur" above the "Bearing Wood," or "Bow," or, as the European vintners have it, "Never put the apprentice above the master," a saying in which lies the whole idea of so trimming as to have the proper number of apprentices ready below, to become subsequent masters. The thighs should never be shorter than eight or ten inches, nor longer than four feet; nor should the bows have more than ten to twelve buds, nor the spurs more than two or three buds. No vine should have more than three thighs—two is enough; nor any one thigh more than one bow, and two spurs, (one will generally be enough.) Should it be intended to get "Layers," it is best to train ground shoots for the purpose.

HOW TO CULTIVATE THE SOIL OF A VINEYARD.

BEFORE any cultivation of the ground, the vineyard should be cleared of all offals from the previous trimming. These offals should be packed down in those spots in the vineyard which have a tendency to wash. They may be used for these purposes in other localities. In Germany, they are gathered up for firewood, as was the old custom in Judea. I have found them most excellent in smoking hams and meat generally, and fancied they gave to meat a better flavor.

The soil of a vineyard should never be cultivated except in dry, warm weather. The drier and warmer, the better. This should especially be the rule in the spring, as ground broken up wet subjects the plants near to injury from even the lightest foot-prints. The rule for breaking ground, in the spring, is simply this: Cultivate as soon as the ground is dry, and warm weather sets in; and don't cultivate, no matter how late you have to wait, until fair weather does set in, and the ground is dry. Don't be in too great a hurry, but improve every fair opportunity. If you delay too long, the buds will swell, and then they are very apt to fall off on even slight shaking.

Whether hoes, ploughs, or cultivators are proper tools, must be determined by circumstances; and they will guide every person having any idea of the cultivation of soil generally. The great point is thoroughness—that is, in turning over every part of the soil, and the most careful attention towards the destruction of all weeds, *particularly around the head of the Vine.* In the spring, the ground should be broken up at least six inches deep, and the rougher the clods are left the better, so that they are fully turned. Rain and warmth will thus penetrate deeper; the ground will wash less, and the clods will be ready to fall entirely to pieces when the second cultivation takes place. The two-pronged hoe, (*karsch,* in German,) is, in my opinion, the best instrument for this purpose. A good shovel-plough may work for the second, and a cultivator for the third operation. The first should take place before the middle of May, or latest, the first of June; the second, as soon after the vines are through blossoming, and the third early in the fall, (but not in the dog-days says an old vintner at my elbow,) a traditionary rule, for which it may be hard to give a scientific reason, but which I found true to my sorrow, in two instances in which I acted counter to the rule. In each case my vineyard lost more or less of its foliage, and all of its rich, green tint.

There are, as yet, no old vineyards in the United States; but it may be well to mention that old vineyards must be cultivated with greater care than young ones; and their roots, even when near the surface, should not be needlessly torn up and injured.

The same care should also be had in the fall cultivation, so as to disturb as little as possible the tender surface-roots, which grow annually out of the head.

The Soil and its Preparation

by

Peter B. Mead

THE SOIL, AND ITS PREPARATION—MANURES.

Soil.—The soil may next occupy our attention. What is the best soil for the grape? This question has been variously answered. Those who live in a district where clay abounds say that a clayey soil is best; while those who live where sand prevails will tell you that a sandy soil is best, and so on. The solution of these answers may be found in the fact that good grapes are grown in both kinds of soil. Our own experience, and a pretty extended observation among vineyards, lead us to give preference to sandy or gravelly loams. It has been said that any soil that will grow good corn will grow good grapes. We have no doubt of the truthfulness of the remark; and we should not hesitate to plant a vineyard upon such a soil, if favorably located. But we may go further, and say that good

The Soil, and Its Preparation.

grapes may be grown where good corn can not. Some of the best vineyards about New-York are planted in light sandy soils, to which muck has been added with a more or less liberal hand. There are many localities on Long Island and in New-Jersey, where light sands prevail, that could be converted into productive vineyards at a comparatively small expense. We have never seen better grapes than have been grown on similar soils properly treated. The vine has such a wonderful power of adaptability that the soil, whether light or heavy, becomes almost a matter of secondary importance.

Preparation.—Not so, however, its preparation for the reception of the plants. This should be most thoroughly done. In planting a vineyard, we are doing a work that is expected to last for generations; hence, every thing connected with it should be done in a manner to insure good and permanent results. Some soils will need more thorough preparation than others; but all will need more or less.

It may, or may not be, that some have recommended a more thorough and expensive mode of preparation than the case calls for. We leave each one at liberty to judge for himself, with the simple remark, that money spent in a judicious preparation of the soil is capital well invested, which is certain to return a good interest. A vineyard well prepared will pay better than one not so prepared: that may be received as an axiom in vineyard culture.

There are three principal methods of preparing the soil for a vineyard: *trenching, trench plowing,* and *subsoiling.* The first, except for small vineyards, and under peculiar circumstances, may be too expensive an operation for general adoption: it is chiefly confined to the garden. The second and third are exceedingly useful, and may be adopted wherever a plow can be run. We propose to give a brief description of each of the three methods above named.

Trenching is done with the spade. It consists in first removing the earth from a trench to the depth that it is proposed to work the soil, the trench to be of any convenient width, (say two feet wide,) and as long as the plot of ground to be trenched. To be a little precise, we will suppose the soil is to be trenched to the usual depth of two feet: the trench will then be two feet deep. With a line, mark off a slice two feet wide immediately adjoining the open trench; throw one foot of the *top soil* of this slice into the *bottom* of the open trench, and on the top of this throw the remaining foot of *bottom* soil. By this operation the trench has been filled, and the order of the soils reversed; the best, or surface soil, being at the bottom of the trench, and the poorest, or subsoil, on the top. We have at the same time opened a new trench. This is to be filled in the same manner as the first, and the operation repeated until the whole plot has been trenched. The last trench is to be filled with the soil that was removed from the first. If the plot of ground is large, some labor will be saved by making the trenches half the width of the plot, going down on one side and returning on the other. The last trench will then be on a line with the first, and there will be but little carting needed to fill it. This is a brief description of trenching, but we hope sufficiently plain to be understood. It will be observed that our operation has buried the good soil, and brought the poor or subsoil to the surface, which must be enriched with muck, manure, or good surface soil from some other place, and we shall have a soil that will bring any kind of plants to their highest state of excellence.

Trench plowing is much less expensive than spade trenching, and but little inferior to it, when well done, putting the ground in fine condition for growing grapes as well as other crops. In trench plowing, oxen are to be preferred to horses, their draught being steadier as well as more powerful. There is no plow in use at present specially adapted to this work, and we must therefore take the best we can get. The cylinder plow, on account of its easy draught, is perhaps one of the best. Two plows and two yokes of oxen are used; the work will be better done, however, if two yokes of oxen are attached to the second or following plow. The first plow opens a furrow as deep as the plow can be driven. The second plow follows immediately in the same furrow, and deepens it to the full capacity of the team. There must be no balks or jumps; the plow must be plunged in to the beam, and kept there. Men with spades should follow the second plow, to remove the stones, and keep the furrow open. The lot may be plowed round, or in lands; but we prefer to return without a furrow, so that the furrows may all be laid one way; the work will be more than enough better to pay for the additional labor. The work will be easier at the start, if both plows are run a second time in the first furrow, and the soil thrown out with spades; the plows will move easier in the subsequent furrows, as there will be less resistance to overcome. A common mistake in trench plowing, (and in all plowing, in fact,) is cutting the furrow slice too wide. It is true, that by cutting the furrow slice twelve inches wide we can get over the ground about twice as fast as when it is cut six inches wide; but in the latter case the work is more than twice as well done; and since we can not do it but once, let us do it well. Let the furrow slices, therefore, be narrow, and the furrows deep. The work will be all the better if the lot is cross-plowed in the same way. The plowing may be repeated with advantage as many times as can be afforded. This would very well meet our idea of *thorough* preparation with the plow. The manures used may be spread on the surface, and plowed in. The effect of trench plowing is not only to deepen the soil, but to mix the surface soil and subsoil together pretty thoroughly, and thus afford a deeper bed for the roots of plants to work in; but among its most important results is the protection it affords against the ill effects of sudden changes of the weather, drought and wetness, heat and cold, etc.

Subsoiling will next be described. This, for the vineyard, is the least thorough of the three methods named. It is but little, if any, less costly than trench plowing, and should not, therefore, except for very good reasons, supersede it. The process of subsoiling is very similar to that of trench plowing. Two plows are used, the common plow and the subsoil plow, which is simply a foot-piece in some wedge-shaped form, attached to a narrow upright shank. Of subsoil plows, there are only two or three in use, either of which will answer the purpose well enough if the furrow slices are made narrow. Mapes's has the lightest draught. In subsoiling, the furrow is opened with the common plow; the subsoil plow follows in the same furrow, and should be run up to the beam to make good work. The lot may be plowed round or in lands; sloping ground, however, should be plowed up and down the slope when the soil is at all heavy;

for the subsoil plow, in such soils, will leave an opening at the bottom of the furrow, which will for a time serve the purpose of a drain. There is this marked difference between subsoiling and trench plowing: the operation of the first is confined chiefly to loosening the subsoil, while the latter not only loosens the subsoil, but mixes it with the upper or surface soil. The value of trenching, trench plowing, and subsoiling, may be taken in the order in which they are named; and it is only the expense of the first which should prevent its general adoption for fruit culture.

Soil and Situation

by

William Chamberlain Strong

SOIL AND SITUATION.

WHEN we consider the exorbitant price of some of the vineyard-lands in favorite localities (some spots on the Rhine being appraised at eight to ten thousand dollars per acre), we might infer that it is only in these localities that we can expect good results. Price is indeed the great index of the comparative value of an article. Applying this index to grape-lands, we shall find a vast preference given to one situation over another. Probably the Rhine vineyards are valued at a higher rate for the mere purpose of cultivation than any other land on the globe. Quite different in character, yet held at the enormous prices of a thousand to five thousand dollars per acre, are the wine-districts of Bordeaux and of Burgundy. In our own country, the same partiality is manifested, to some degree, for favorite localities. For example, some lands bordering upon Lake Erie are held at from two to three hundred dollars per acre and upwards, which is a great increase over ordinary farming lands.

A brief description of the peculiarities of the most famous European districts will be interesting and suggestive. The world-renowned vineyards of the Rhine district are planted on both sides of the river, some of the most famous having even a due-north aspect. They are described as having a good deal of clay mixed with the loose stony soil. When a vineyard becomes exhausted after a culture of about thirty years on these steep slopes, it is renewed by adding several inches of clay as well as manure. The clay is necessary to give strength to the otherwise gravelly and loose stony soil. The Steinberg lands are a bluish clay, the substratum being gravelly. Most of the Rhine soil, the famous Johannisberg for example, is a very red clay, with gravel freely intermixed. In the Burgundy district, the finest wines are produced from vineyards upon the *Côte d'Or* (Golden Hills). This range stretches from Chalons to Dijon, a distance of eighty miles, in a north-east and south-west course. The soil is described as red and gravelly, containing a good deal of limestone. At the top (an elevation of two to three hundred feet) there is but little soil, the red rock projecting in many places. The vineyards commence nearly at the top, where the soil is reddest, and where the richest wines are produced from the small black Pineau Grape. The middle range of the hills is planted with the Black Gamai, larger and more prolific, but yielding an inferior wine. Third-class vineyards are planted down to the foot, and extending into the plain, producing abundantly, but giving only ordinary wine. The hills on both sides of the River Marne are planted; but the sides looking due south are classed differently from those looking north. The south slopes include such distinguished vineyards as Hautvilliers, Disy, and Aix; while the equally famous Epernay, Moussy, and Vinay are on the opposite bank, looking north. The Mountain of Rheims, though in the north of France, is planted on its northern as much as on its southern slope; the soil being a limestone and chalk formation, with a thin covering of vegetable matter. Among the Pyrenees, the vineyards are extended half-way up the highest mountains. In a comparison of French and Hungarian wines, M. de Szemere writes as follows: "In Hungary the old system prevails, under which the quality is the principal object in view, and under which a favorable exposure is the all-important consideration; and the poor, light, stony, granitic land, from whence alone the choicest and the most highly-flavored wines can be obtained, is preferred to a rich, manured soil which insures an abundant, but, in quality, far inferior return. Nothing is grander or more beautiful than our mountains, crowned either with shady woods, or with vines of exuberant vegetation. Where you see a mountain, there you will find our vineyards. The superb Badacsong Mountains form a high semicircle around the majestic Lake of Balaton, covering a surface of a hundred and twenty-five English square miles. The arid mountains of Ménes or Vilàgos overlook proudly the rich plains of Bànat, the holy Canaan of Hungary. The mountain called Tokay rises in another large plain like a lofty pyramid. It has the form of Vesuvius, and, indeed, its existing but silent crater: its volcanic formation shows evidently that it was once a fire-spreading mountain. The cultivation of such a soil is very difficult and expensive, the produce obtained but little; but then the latent fire of this volcanic mountain is what we call Tokay wine."

The above examples confirm the truth of Virgil's oft-quoted statement, "Bacchus amat colles." Yet we find marked exceptions to this rule in various parts of Italy,

France, and other countries. Falernia, whose wines were so celebrated in classic song, was a fertile plain. The Medoc district, near Bordeaux, is a gently undulating plain, extending from the River Gironde on the east to the Atlantic on the west, with frequent lagoons indenting the shores on either side. This peninsula contains some of the finest vineyards in the world, such as Lafitte, Château-Margaux, Branc-Mouton, &c. The soil is a coarse, sandy clay, strongly impregnated with oxide of iron. The vineyards of Languedoc, of Tonnere, and on the banks of the Rhone, are of this level or slightly undulating character. Although a chemical analysis of a soil is a very uncertain guide, independent of other conditions, yet the following table of the soil of the celebrated plain of Château-Margaux will be of interest :—

>   Oxide of Iron 3.341
>   Alumina................................... 1.590
>   Magnesia 0.263
>   Soluble Silicates........................... 0.380
>   Phosphoric Acid............................ 0.147
>   Potash 1.291
>   Carbonate of Lime......................... 0.891
>   Organic matter 6.670
>   Insoluble residuum........................85.427
>   100.000

It cannot be doubted, that, with a favoring climate and soil, an excellent wine can be obtained from plain lands. We know, that, under French manipulations, these wines become famous; yet it is equally certain that the very highest wines can be obtained only where the growth and produce have been quite limited, and the fruit has attained the most concentrated flavor from an abundant amount of light, air, and heat.

There is no apparent reason why the rule which is observed in Europe should not hold good in this country. Certain conclusions may be drawn from a study and comparison of European methods. We may conclude that the production of grapes for the table and for wine are two distinct purposes. For the table, we require fair, large, and luscious fruit, full of juice, bunches of good size and form, an ornamental as well as a useful fruit. These conditions require a generous growth, which will give a large supply of watery matter at the expense of the high saccharine and vinous quality which is so much prized by connoisseurs. A good home drink can undoubtedly be made even from the gleanings of our plain vineyards; but if there is any aim to produce wines which will bear the test of comparison with those of the Rhine, of Burgundy, or Tokay, we must learn to account

7

quantity as of least importance, while quality is the *sine quâ non*. Again: it will be noticed that the hillsides flanking a river are universally esteemed. The deeper these valleys, the farther north the culture of the vine may be successfully extended. This is the secret of success in the Rhine Valley, the grape being planted even upon the northern slopes of this high latitude of fifty-one degrees. In these valleys the air becomes heated during the summer much more so than in the open plain. In addition, a moderate humidity is preserved by the mist arising from the flowing river. It will be noticed that the Rhine lands are described as having a large proportion of clay. It should be borne in mind that this is freely intermixed with calcareous and silicious gravel, while the sharp pitch of the hill will insure quick drainage and a warm soil. As a rule, we observe that preference is given to a loose, warm soil; limestone and silex being both considered as desirable elements. On the plains, more sand is required than on the hills: a heavy clay or loam, at all inclined to dampness, is unfavorable. Sufficient clay to give strength to a dry, gravelly hillside, or a sandy plain, would be a valuable addition. Just that degree of richness should be sought that will insure health and a fair degree of vigor, in order to the full development and early maturity of the fruit; at the same time guarding against such fat lands as will produce rank and immature growth. Some varieties of the grape (*e.g.* the Rebecca) require more strength in the soil, more clay, than others; but this should never so abound as to be called heavy land.

In respect to the aspect of hill-slopes, the testimony of the majority both in this country and in Europe is decidedly in favor of a southern exposure. An eastern exposure is good, as having the benefit of the early sun; a south-east aspect being still better, as receiving the warmth for a longer time. A western slope is shaded from the morning sun, an hour of which is reckoned by gardeners to be worth two hours of the evening sun. After the cool moisture of night, all plants long for the warm rays of early day. A northern slope is oftentimes so situated as to receive the sun's rays from rising to setting. Such aspects may be allowable in many cases, especially for early kinds, and in portions of the country where the fruit has abundant time to mature; yet it must be for other reasons, and not because the slope is desirable, that this aspect is chosen. An exposure to sweeping winds is objectionable, causing much more rapid evaporation from the expansive foliage, and thereby exhausting the life of the vine in hot, dry weather. The contrast between such arid exposure and the comparatively moist and confined heat of river hillsides is very perceptible on the vine. It is found also that vines greatly exposed are more liable to mildew than in positions where a more uniform humidity is preserved and the changes are less violent. It is well known, that, in many parts of the country, the grape does not ripen as well as in former years. In Massachusetts, the Isabella, for example, used to be a certain fruit, but has now become almost a certain failure in ordinary localities. Our State was formerly covered with forests, — Nature's vast system of sponges, — which absorbed the rains, and gradually gave them off in the humid atmosphere, and in gently flowing streams, for months afterwards : but now the greater portion of the country is laid bare to the fierce rays of our clear sun; the natural mulching of leaves is lost; our rains rush in torrents down our hillsides, and speedily make for the ocean whence they came. By this we do not imply that our climate can, with any propriety, be called arid; yet it is true that there is much less *uniform* humidity of the atmosphere during the summer than in former years. An able presentation of this subject will be found in the volume of Hon. G. P. Marsh, entitled "Man and Nature," to which the reader is referred.

The drainage of our meadows and bogs will have an influence in the same direction; so that, by the slow but constant effort of man, silent yet vast changes are effected in the entire system of Nature. This diminution of the even humidity of the air during summer, and the increase of strong sweeping winds with more sudden and violent changes, we cannot doubt, is prejudicial to the grape. We must resort to means to counteract this evil; and, as has been said, we must avoid arid positions exposed to strong currents of wind; seeking protected situations having, if possible, an evaporating surface of water near at hand. Not that more moisture is needed, but that the modifying influence of the lake or ocean may give more uniformity to the humidity and the temperature. Hence the shores of Lake Erie, the Hudson, the Rhine, the Rhone, and the Bordeaux peninsula between the Gironde River and the Atlantic, are all highly prized. It is within the power of man greatly to modify the character of a situation by the judicious planting of belts of evergreens, by a wise addition of elements and a proper culture of the soil, and by encouraging the shade of the vineyard itself wherever circumstances indicate its necessity.

The mechanical texture of the soil is perhaps of more consequence than its chemical analysis. It should be loose and friable; limestone and silicious sand being always esteemed desirable elements. Limestone soils are found to absorb more of the sun's heat during the day, and to part with this heat much more slowly at night, than is the case with vegetable soils. The same is true also of silicious soils; so that they may with propriety be termed warm soils. In conclusion, we may add that the vine will exist in almost any variety of soil; that it luxuriates in rich, fat lands, the growth of wood being excessive, and the fruit large, well developed, but lacking in quality; and that in lighter and dryer soils the growth and fruit are less, but the quality is superior, and the plant is much less liable to disease. A common and rough, yet in general a correct test, is found in the question, whether a soil is favorable for the growth of Indian corn.

PREPARATION OF THE SOIL.

Many soils well situated, and naturally adapted to the grape, are yet wet and springy. It is of great importance that the subsoil as well as the surface should be thoroughly drained. It must be determined, before planting, whether this work is necessary. The distances for the drains will vary from twenty to forty feet apart, according to the nature of the soil. It is desirable to place the tile from three to four feet deep, in order that they may be out of reach of the roots. We are next to consider the nature of the soil, the purpose for which the grapes are cultivated, and the particular kinds to be planted. If the soil is a light, silicious sand, some clay may be added with other enriching material. A strong gravelly soil will require a compost of two-thirds vegetable mould with one-third of stable manure which has been mixed for some time previous.

Fifteen cords of this compost to the acre is little enough dressing for most New-England soils. Indeed, for the purpose of obtaining table fruit, and for such varieties as the Delaware, this quantity may be doubled; but, for such strong and long-jointed kinds as the Concord, a less amount than that first named will generally be found sufficient. The compost is to be spread evenly over the whole surface. To this compost we may add, with profit to most soils, at the rate of two barrels of unslacked lime for every acre. Wood-ashes are always in order for the grape: yet the sole object for the first two years is to produce wood only; and for this the natural strength of the soil, aided by the compost manure, should be fully equal. The bulk of ashes, bones, lime, sulphur, phosphates, or other special manures, should be reserved for top-dressing in the third year and thereafter. We are now ready for the work of loosening the soil and incorporating the compost. Trenching is recommended by some, working the entire surface with the spade to a depth of two feet at least. Some authors recommend that the subsoil should be brought to the top, and the surface should go to the bottom; others would preserve them in their relative position, simply loosening the earth; others still advise that the two soils should be well mixed. Different soils require a modification of every general rule; yet the latter course seems to be most reasonable, and has the approval of most practical cultivators. But this handwork, though most thorough and perfect, seems to me to be too laborious and expensive for any extensive application. In most soils suited for a vineyard, the plough can be made to go to the depth of twenty inches or two feet, and the work is done sufficiently well for practical purposes. The first furrow being opened as deep as is possible with a strong team, let the subsoil plough follow in the furrow, not only loosening the subsoil to the required depth, but also mixing the surface soil to an extent, as I think, sufficient for all practical purposes. On the hillside, the side-hill plough will throw the surface furrow so far down the hill as to give full space to work the subsoil. This work of ploughing should be done in the fall. If, however, the compost is not in readiness in the fall, it may be spread in the spring, and worked in by a second surface-ploughing.

In case the situation is a hillside, the question of terracing will arise. Undoubtedly this may be advantageous in some cases. Narrow terraces, supported by a stone wall, or even a grass sod, will prevent wash, and give the vine a warm and protected exposure; but this extra expense will not be warranted except the situation is very steep, so that the wash will prove excessive. A partial terracing may be done without great expense, provided a sufficient quantity of stones, from one to two feet in diameter, is at hand. A single course of such stones running in parallels along the face of the hills, and just behind each row of vines, will be a great help in preventing wash. On many hillsides, it would not prove tedious, or very expensive, to throw narrow terraces when the greater part of the work can be done with the plough, the bank and other finishing work being left for the shovel. In such a case, the top surface of the terrace should slope slightly towards the hill, in order that rains may not run down the bank, and wash. Where a more expensive system of terracing is adopted, it will be rather for ornament than for profit; and this may be left to the landscape-gardener.

VINE-BORDERS.

We have thus far spoken only of vineyard-lands. Many persons will wish for instructions for borders for a few vines exclusively for table-fruit. In such cases, the borders may be made deeper and richer. We have instances where the vine is an exceedingly gross feeder. The Hampton-Court Vine lives upon the sewerage of London. We read of artificial borders, three to four feet deep, one-third of which is rich stable-manure, with a large supply of bones, lime, &c. The result is a prodigious crop of grapes. We are now speaking of foreign varieties, under glass, which receive constant and peculiar care. Let it be remembered also that such excessive growth is only of second quality for the table, and would be utterly worthless for wine. There is a limit to the richness of a border; though the American people will incline to err on the side of extra growth, to the sacrifice of quality. In such rich borders, another difficulty occurs: the vine is inclined to make wood rather than fruit. This is much more true of some varieties, *e.g.* the Concord, than of others. The Delaware, and such other kinds as are short-jointed, prolific, and of moderate growth, will allow, and even require, a rich soil. If, then, the purpose is to give special advantages to one or more vines for table-fruit, as the first step, see that the sub-soil is well drained. Then add a sufficient quantity of friable pasture sod to increase the depth of surface-loam to eighteen inches: to this add six inches of stable-manure, and about a bushel of bone-dust, to a square rod of border. If the soil is inclined to be heavy and retentive, add pure sand; or, if sand is in excess, add a moderate quantity of clay, and perhaps well-decomposed vegetable soil. But peat and meadow-muck are not desirable additions to such a soil as may be called a good garden or pasture loam. To such a loam, six inches of manure will be quite a sufficient supply of vegetable matter. In such rich borders, special care is necessary that they do not become too heavy, and retentive of moisture.

An addition of lime will be beneficial in correcting this tendency, and will also be of service, both as food for the plant, and as assisting in the decomposition of organic matter and in destroying insects. Potash, in some form, is essential to the grape; yet this may better be supplied as a surface-dressing during the after-growth, when, as we shall see, it is most needed in producing fruit.

The border must be trenched, mixing all the materials, simply loosening the sub-soil, and letting it remain at the bottom. If, however, the character of the sub-soil should be judged suitable to have a good mechanical effect upon the surface-soil, or if it contain any elements which are desirable on the surface, it should be brought up, and mixed to a greater or less extent. The border will be two feet deep when finished, deep enough and rich enough for the feeblest variety; too much so for rampant kinds.

Some authors recommend the grossest and richest materials for the border, such as slaughter-house offal, whole carcasses of horses and cattle, and the like. This is with the object of giving permanence to the border, it being supposed that these remain a store of fertilizing wealth for many years; the large bones becoming fit for use as the fleshy matter is exhausted. Provided such gross material is buried to such a depth that the young roots do not come in contact with it in its putrid state, it may do no present injury; but it will ultimately draw the roots to such a depth from air and warmth as to prove objectionable.

Such use of material is not only a waste, but a positive injury. The use of whole bones is a decided benefit, yet a most expensive mode of applying nourishment. Some kinds of bones will remain in the ground for half a century with but little change. It is manifest that it would be poor economy to furnish so expensive and valuable food by such a slow process. It is true that vine-roots will intwine themselves around and through fresh bones, and greedily take up whatever is obtainable; yet the great bulk is forbidden food until crumbled by time. It is said that vine-borders should be supplied with permanent material which should last as long as the vines themselves. It is indeed of the utmost importance that the composition of the soil should be such that it shall never become sodden, or suffer from drought. Being originally composed of suitable earths, and a moderate supply of organic matter, it will be easy to add manures on the surface, from year to year, as the land may require. Unquestionably the most economical mode of applying manure is to reduce it, as nearly as possible, to a condition for immediate use as food for the plant. Stable-manure should be well composted and decomposed. Bones should be broken into fine pieces, or, better still, ground to powder. Nothing is lost by this process, and a great gain is obvious, both in time, and in other respects which it is not necessary to enumerate. If, then, we can answer four requests, — viz., thorough drainage, a friable soil, a generous dressing of composted manure, and loosening to a depth varying from eighteen to thirty inches, — we shall be ready to take the next step of planting.

Soil and Situation

by

Andrew S. Fuller

SOIL AND SITUATION.

WHEN we take into consideration the wide extent of territory in which the grape is found growing, either in its wild or cultivated state, on both the Eastern and Western continents, we may well ponder over the oft-repeated assertion, that the vine does not succeed over the whole extent of any country, but only in certain circumscribed localities; and while we may admit its truth, we fail to comprehend the reasons why certain soils or sections of a country should be more congenial to the vine than others. Yet the fact that success does attend its cultivation in particular locations, while it entirely fails in others, is patent to every casual observer. Whether these failures are attributable to the injudicious selection of varieties, or to the mode of culture, is not always easy to determine. That the climate of both the Northern and Southern States, as well as that portion of the United States lying west of the Rocky Mountains, is congenial to the vine, is abundantly proved by the numerous varieties found growing wild over this vast region of country. No doubt there are particular varieties which are better suited to one section than to another, and that the same situation and exposure that would be most suitable for a vineyard at the South, might be the worst that could be selected for the North.

A situation protected from the cold north winds, so as to insure sufficient heat to mature the fruit, is always desirable in a cold climate; but in a hot one the heat may be so great as to exhaust the powers of the vine by a too rapid evaporation from its leaves, and it consequently fails.

SOIL AND SITUATION.

Nearly all the writers on grape culture recommend the declivities of hills and mountains inclining to the south as the best exposure for a vineyard; and the next in order are the southeast, east, southwest, but never a north or a full western exposure. Virgil said, "Nor let thy vineyard bend toward the sun when setting," and these words are as applicable at the present time as they were two thousand years ago.

A full southern exposure is no doubt preferable in the Northern States, and if the land descends to the south, so much the better; but if very steep, it will cost more to prepare and keep in order than if it is level. While I admit that a side-hill is a very desirable location for a vineyard, I am quite certain that there are many situations equally good that are perfectly level or nearly so.

I have often observed that the success of a vineyardist was attributed to his soil and situation, but never to the skill of the cultivator or to the varieties grown, and this, no matter whether his soil was light or heavy, or the situation low or elevated. Still, we know that soil and situation have often much influence upon the growth and quality of the fruit; for the instances of such an effect being produced are too common in Europe, at least, to allow us to deny its truth.

In selecting a situation for a vineyard, all the surroundings should be closely observed and taken into account. If the land has no protection from the north and northwest, see what the facilities are for supplying one either by walls or a belt of trees. [If trees are to be used, evergreens are best, and often the small trees may be had in the woods near by—we now refer to the Northern States.] See that the land is sufficiently elevated, thirty to forty feet at least above streams or ponds of water; for, if near the level of small bodies of water, the situation will very likely be subject to early and late frosts.

Large bodies of water are not so injurious as small, as

they absorb heat in such quantities during summer and give it off slowly in the fall; this affects the surrounding country very materially by preventing early frosts. In spring, the water being cold, it keeps the atmosphere cool for quite a distance from the shore, and thereby prevents vegetation starting so early as it otherwise would.

This, I have always observed, was the case on Long Island; we seldom have frost as early in the fall as they do a few miles back on the mainland, and vegetation does not come forward so early in spring as to be cut off by late frosts.

When the soil is sandy or gravelly, it will require an application of some organic materials, either in the form of barnyard manure, muck, or leaf-mold. The latter two can often be readily obtained, where the former, in any considerable quantity, would be out of the question, or would be so expensive that it would very much lessen, if not entirely absorb, the profits of the vineyard. There are thousands of acres of sandy or gravelly lands in the Eastern States that would make the very best vineyards in the country, simply by applying the enriching materials that are to be found in abundance in their immediate vicinity.

Strange to say, these lands are now considered almost worthless, because barnyard or special manures (as they are called) can not be had sufficiently cheap to make them profitable for cultivation. While a sandy soil may not naturally produce the most luxuriant growth, it is certain that it produces fruit of the richest quality. Such soils are moderately favorable to the growth of the vine, are easily worked, and do not retain an excess of moisture, as they are thoroughly underdrained by nature.

Both granitic and limestone soils are excellent for the vine, and as they are usually what is termed strong soil, they need but little more than a slight change in their mechanical condition, which is readily accomplished by plowing or trenching. And here let me remark, that very often the mechanical texture of the soil has more to do with success or failure than do the ingredients it contains.

A moderately loose and friable soil, whether it be loam, sand, gravel, or the debris of rocky hillsides, especially if of a calcareous nature, are to be chosen in preference to clay or muck. These latter may be somewhat reclaimed and made available by underdraining, trenching, etc., yet in a majority of cases they prove unsatisfactory in the end.

The soils in many portions of the Western States, and in some portions of the others that have but recently been brought under cultivation, need no addition of fertilizing materials.

New soils are to be preferred to those that have long been in cultivation; for it is extremely difficult to supply artificially to worn-out soils the lacking materials in a form so perfectly adapted to the wants of plants as that which they originally possessed. I am well aware that some agricultural chemists have endeavored to impress upon the minds of cultivators the importance of analyzing the soil, in order to ascertain what particular ingredients it may need, or what it may possess in too great an abundance to produce any particular crop or plant in perfection. And while I admit that chemists may sometimes determine when there is an excess of any particular constituent (which practical men will often do by merely looking at it), I have yet to learn that analytical chemists can tell how little of any particular ingredient is needed for any particular crop. An acre contains 43,560 square feet of surface, and if we call the soil a foot deep (and there are few plants that do not penetrate deeper than this), then there will be that number of cubic feet. A cube foot of ordinary soil will weigh from 75 to 100 pounds—we will call it 80 pounds—this gives 3,484,800 as the weight of an acre of soil one foot deep. There are circumstances of frequent occurrence when a farmer, by adding 100 pounds of some particular material to an acre of grain, will increase the crop twenty-five per cent. And certainly it is not reasonable to suppose, nor do I think that any theorist will maintain, that it is among the possibilities of chemical science to detect even a trace of 100 pounds of a substance in 3,484,800, yet plants will detect it.

I make these remarks because I have seen men, when looking for a situation on which to plant a vineyard, who were very particular to have the soil analyzed by some celebrated chemist before they would purchase or plant. I do not wish to depreciate the science of agricultural chemistry, for it has been one of the powerful auxiliaries in the advancement of agriculture, but I would warn those who implicitly rely on all the theories advanced, that they may ask too much of it, and thereby be led astray.

PREPARING THE SOIL:

The manner of preparing the ground for a vineyard depends entirely upon the kind of soil and its natural condition. If it is heavy and compact, the first step will be to underdrain it either with stone drains or tile.

The number of drains required, and the depth to which they should be laid, will also depend somewhat upon the nature of the soil and the amount of water to be carried off. If the land has springs in it, then the drains should be placed so as to cut them off near their source and prevent, as much as possible, the excess of water from spreading.

But soils containing springs, except they be gravelly, should be avoided, as they are perhaps the most unsuitable that could be selected. There are also many soils that are called dry that should be underdrained, especially if they are inclined to heavy loam or clay, not so much to carry off the water, but to aerate the soil—that is, allow the air to penetrate and circulate through it; for air always carries with it more or less heat and moisture, and if the mechanical texture of the soil is such as to readily admit air, then

SOIL AND SITUATION.

it will be more likely to be in a condition to transmit moisture rapidly, but not to hold a superabundance.

Drains are usually placed from 20 to 40 feet apart, and three to four feet deep, according to soils, situation, and the crop to be grown on the land. For vines, the drains should be placed deeper than for ordinary farm crops, else the roots will soon penetrate to and fill them. To describe the different kinds of materials used in draining lands, as well as the manner of laying, cost, etc., would occupy too much of our space, and we must refer those of our readers who wish to plant a vineyard upon soils that require draining, to those works that treat particularly on this subject.

When vines are to be planted upon steep hillsides or upon stony soils, the only thorough method of preparing the soil is by trenching. This is done by digging across the field to be planted a trench two feet wide and two feet deep—some recommend three feet or more; but if it is full two feet it will generally be deep enough, and deeper than nine out of ten do actually trench when they say three feet. After the soil has been thrown out upon one side of the trench, a parallel strip of soil, of the same width of the trench, is thrown into it, and by this means the soil is inverted, the top or surface soil being placed at the bottom, and in this way one trench is dug to fill up another, until the whole field is trenched over. The soil taken from the first trench will consequently remain on the surface above the level of the surrounding soil, and there is no soil to fill the trench last made. It is usual, on level ground, to take the soil that was dug out from the first trench and put it in the last; but to do this is often inconvenient, and I have yet to see a piece of land, of any considerable size, without a spot somewhere upon it where the soil thrown out of the first trench would not improve it by filling it up; and if the trenching is finished off upon the higher portion of the field, the trench last made may be filled up from the adjoining soil without injuring its looks. It does not matter where we begin to trench, whether in the middle or at one side of the field.

This inverting the soil, as described, is the simplest method of trenching, and is as efficient as any, provided the subsoil is not of a character so inferior that it will not be rendered suited to the growth of plants by being exposed a few months to the atmosphere. The subsoils of light sandy soils are often richer than the surface, as a great portion of the enriching materials that have been applied to the surface has been carried down by the rains to the subsoil below. But the natural richness of the subsoil when thrown upon the surface should seldom be depended upon, but manure must be added, and thoroughly incorporated with it before planting.

There are many circumstances in which the soil may be inverted to the depth of two feet with benefit—such as sandy or light loamy soils, or where manure can be liberally applied, and a year be given for the amelioration of the condition of the subsoil before planting. Where these circumstances do not exist, it is best not to throw the subsoil on top, but to mix it with the surface-soil at the time of trenching.

To do this thoroughly and conveniently, the workmen should stand in the trench, and keep an open space at the bottom on which to stand. Then, by digging down the bank and throwing it over against the opposite side, break up the lumps at the same time; the soil may not only be thoroughly pulverized, but the surface-soil and subsoil will become thoroughly intermingled.

A five-tined spading-fork is the best implement for trenching unless the soil is very hard and stony, in such cases the spade and pick must be used.

Although trenching is indispensable upon very hard and stony soils and upon steep side-hills, on level situations or those with only a moderate inclination (and they are always preferable) the plow may take the place of the spade, and it will very much lessen the expense of preparation. If the soil is stirred to the depth of twenty inches, which it may be by using a subsoil plow, it will be as deep as is really necessary to insure a good and healthy growth of vine.

I do not believe that it is desirable to encourage the roots to penetrate to a great depth, especially in a northern climate, for when the roots penetrate deeply they do not come into full action until late in the season, and they also continue to act late in the autumn, thus preventing the early ripening of the wood.

If the soil is not naturally rich, spread the manure upon the surface before plowing, then turn it under with the surface plow, and let the subsoil plow follow in the same furrow, breaking up the subsoil. After the ground has been all plowed over in this way, then cross-plow it in the same manner; this will insure a thorough breaking up of the soil and mixing of the manure with it. When the land has been both plowed and cross-plowed, if it is then gone over with the cultivator it will still benefit it very much, as it will break to pieces the lumps which will be left unbroken after even the most thorough plowing.

We should always endeavor to make thorough work in the preparation of the soil before planting the vine, for it is not an ordinary crop that we are to plant, nor one that necessitates a seed-time to each harvest, but it is one that requires but one planting in a lifetime, yet it will reward us with many harvests.

There are very few soils that a person of good judgment will select that will need any further preparation than that which can be done with the plow, with the addition, perhaps, of underdraining. Manures, of course, must be applied where the soil is not rich enough without them.

www.ingramcontent.com/pod-product-compliance
Lightning Source LLC
Chambersburg PA
CBHW031316150426
43191CB00005B/255